Poverty and the Third Way

T0264493

What is poverty and how can it be tackled? The Old Left adopts redistribution as the solution to poverty; the New Right identifies the poor as an underclass in need of integration into the mainstream; and New Labour's 'third way' sees the root cause of poverty as joblessness and integration into the formal labour market as the solution. Taking the 'third way' out of its narrow party political context, this book argues that a new path beyond capitalism and socialism can only be paved if the third way frees itself from its obsession with employment.

Poverty and the Third Way uncovers how New Labour's employment-focused approach causes, rather than resolves, poverty. Searching for an alternative 'third way', the authors find the seeds of such an approach in radical European social democratic and ecological thought that seeks to transcend capitalism and socialism by developing work beyond employment. Exploring the rationales for such an approach and how it can be implemented, the authors transcend the mindset that views there to be 'no alternative' to capitalism, by providing a clearly marked route map of the way towards a post-capitalist economy.

This book will be of interest to academics and advanced students within the disciplines of human geography, social policy, sociology and economics, as well as offering great value to practitioners in the fields of community and economic development, and to those with an interest in central government and supra-national policy-making.

Colin C. Williams is Reader in Economic Geography at the University of Leicester. **Jan Windebank** is Senior Lecturer in French Studies and Associate Fellow of the Political Economy Research Centre (PERC) at the University of Sheffield.

Routledge Studies in Human Geography

This series provides a forum for innovative, vibrant, and critical debate within Human Geography. Titles will reflect the wealth of research which is taking place in this diverse and ever-expanding field.

Contributions will be drawn from the main sub-disciplines and from innovative areas of work which have no particular sub-disciplinary allegiances.

Poverty and the
Third Way

Colin C. Williams and
Jan Windebank

Routledge
Taylor & Francis Group

LONDON AND NEW YORK

To Toby for being so brave

First published 2003 by Routledge
2 Park Square, Milton Park, Abingdon, Oxfordshire OX14 4RN

Simultaneously published in the US and Canada
by Routledge
711 Third Avenue, New York, NY 10017

First issued in paperback 2015

Routledge is an imprint of the Taylor and Francis Group, an informa business

Typeset in Garamond 3 by
Florence Production Ltd, Stoodleigh, Devon

British Library Cataloguing in Publication Data
A catalogue record for this book is available from the British Library

Library of Congress Cataloging in Publication Data
Williams, Colin C., 1961-
   Poverty and the third way / Colin C. Williams and Jan Windebank.
      p. cm.
   Includes bibliographical references and index.
   ISBN 0–415–25725–5 (alk. paper)
      1. Informal sector (Economics) 2. Mixed economy. 3. Mutualism.
4. Poverty. I. Windebank, J. (Janice) II. Title
HD2341 .W485 2003
339.4′6–dc21   2002068179

ISBN 13: 978-1-138-88336-9 (pbk)
ISBN 13: 978-0-415-25725-1 (hbk)

Contents

Tables

Acknowledgements

This book reports the findings of five separate but interrelated research projects conducted over four years. The data set on the coping capabilities and practices of households in urban and rural England is the cumulative result of three sources of funding. The Joseph Rowntree Foundation funded the study of urban lower-income neighbourhoods as part of its programme of research and innovative development projects, which it hopes will be of value to policy-makers and practitioners. The facts presented and views expressed in this book, however, are those of the authors and not necessarily those of the Foundation. The authors would also like to thank Stephen Hughes who provided the research assistance to bring this study to fruition.

The data set on higher-income urban neighbourhoods, meanwhile, was funded in part by the European Commission's DG12 under its Targeted Socio-Economic Research programme. This data was collected as part of a project entitled 'Inclusion through Participation' (INPART). In this regard, the authors would like to thank Rik van Berkel who managed the six-nation project team as well as our other European partners. In particular, we would like to thank our UK colleagues, namely Maurice Roche and Jo Cooke, without whom this data could not have been collected.

The third data set on rural localities arose out of funding by The Countryside Agency. We would like to take this opportunity to thank them for funding this project as well as to display our gratitude to Richard White and Theresa Aldridge for their energetic and enthusiastic research assistance.

Many of the ideas developed in this book were first tried and tested in the context of a twelve-nation project funded by the European Commission's Targeted Socio-Economic Research programme entitled 'Comparative Social Inclusion Policy and Citizenship in Europe: towards a new social model'. The regular and enjoyable meetings with our partners from other European Union nations enabled us to forge our ideas in a way that would not otherwise have been possible. In this regard, we would like to thank the following people for their inputs: Claire Ainesley, Rik van Berkel, Soledad Garcia, Henning Hansen, Pedro Hespanha, Iver Hornemann-Møller, Angelika Kofler, Jens Lind, Enzo Mingione, Maurice Roche, David Smith, Ben Valkenburg, Jacques Vilrokx and Enid Wistrich. Although they will doubtless take issue with many of the ideas in this book, their interventions and inputs have played a large part in helping us develop the ideas in this volume.

The fifth and final source of funding that has fuelled the ideas herein was the Economic and Social Research Council (ESRC) grant 'Evaluating Local Exchange and Trading Schemes as a tool for combating social exclusion' (R000237208). This was carried out in collaboration with Theresa Aldridge, Roger Lee, Andrew Leyshon, Nigel Thrift and Jane Tooke. Colin Williams would like to thank all of his co-partners in this project as well as the respondents.

As always, however, the normal disclaimers apply.

Introduction

What is poverty and how is it to be tackled? In a recent widely cited text, Levitas (1998) outlines three discourses that provide very different answers to this question. First, there is the discourse of the Old Left that views the poor to have 'no money' and adopts redistribution as the solution. Second, there is the view of the New Right that identifies the problem as an underclass that has 'no morals' which needs to be brought into the mainstream. Third and finally, there is the approach of New Labour that pinpoints how the poor have 'no employment' and views integration into the formal labour market as the route out of poverty.

Based on this heuristic framework of the discourses on tackling poverty, it might be thought that there is only one 'third way' beyond the Old Left (the 'first way') and the New Right (the 'second way') and it is New Labour's employment-focused approach. In this book, however, our intention is to articulate and develop another third way discourse that, although often only briefly sketched out, is beginning to gain greater recognition as a potential way forward. This discourse rejects the infatuation of New Labour with inserting people into employment as a means of tackling their poverty (e.g. Beck 2000, Giddens 1998, 2000, 2002, Jordan 1998, Levitas 1998, Lister 2000). Instead, the basic premise of this alternative third way approach is that it is the 'capabilities' of people to meet their needs and desires that matters and that if these are to be enhanced, more is required than merely providing them with employment. For poverty to be alleviated and a more inclusive society created, the assertion is that the abilities of people to participate in work beyond employment need to be cultivated so as to enable them to develop their capacities to help themselves and others.

To see the emergence of this discourse, one needs look no further than Levitas (1998) herself. One of the key criticisms that she targets at New Labour's third way is that its employment-centred discourse fails to valorise unpaid work and that unless this is done, then poverty and social exclusion cannot be tackled. Yet although, or perhaps because this thesis is part of her own beliefs, she fails to incorporate it into her classificatory schema as a fourth discourse. If she were a lone voice in the wilderness advocating such an alternative third way approach, then this omission would be understandable.

A cursory glance at other prominent scholars of the third way, however, suggests that this is not the case.

Lister (2000), for example, one of the most prominent social policy analysts in the UK at present, argues that the 'social inclusion through employment' discourse of New Labour is founded upon four assumptions that need to be unpacked and critically appraised. These are that: paid work necessarily spells social inclusion; joblessness necessarily spells social exclusion; the only form of work of value to society is paid work; and an inclusive society can be built on the foundations of paid employment alone. For her, insertion into employment often leads not to social inclusion but to social disintegration (e.g. due to long working hours and home/work pressures) and unemployment does not always lead to social exclusion. Similar to Levitas (1998), her belief is that we need to move beyond the view that paid employment is the only form of work of value to society. To build a more inclusive society, there is a need to recognise and value unpaid work.

In a similar vein, Bill Jordan, a widely respected critic of current New Labour third way thinking, has a recurring and dominant theme running through his prolific output (e.g. Jordan 1998, Jordan and Jordan 2000, Jordan *et al.* 2000). He is highly critical of New Labour's third way approach that equates social exclusion with unemployment and social inclusion with employment. For him, this cannot last. Limits exist with regard to the extent to which formal employment and comprehensive formal welfare provision can be used to tackle the poverty problem. He develops a convincing case (with a touch of teleological inevitability thrown in) as to why New Labour cannot continue to ignore work beyond employment.

Perhaps the clearest and fullest elucidation of this alternative third way discourse, however, is to be found in the scholarly body of work of the principal architect of New Labour's third way, namely Giddens (1998, 2000, 2002). For him, it is the capabilities of people that matter and as such, we need to move away from uni-dimensional employment-focused solutions to poverty and harness people's abilities to help themselves and others. As he explains,

> [Can we] return to full employment in the usual sense – enough good jobs to go around for everyone who wants one? No one knows, but it seems unlikely . . . Since no one can say whether or not global capitalism will in future generate sufficient work [i.e. employment], it would be foolish to proceed as though it will.
>
> (Giddens 1998: 126)

For him, a constituent component of any third way approach is the cultivation of what he calls 'civil society'. Poverty is not about 'no income' or 'no morals' as in the discourses of the Old Left and New Right (see Levitas 1998). Nor, moreover, is it solely about 'no employment' as in the social integrationist discourses of New Labour. Instead, and following the work of the

Nobel prize-winning economist, Sen (1998), Giddens (2002: 39) argues that it is 'people's capacities to pursue their well-being' that matter and as such, we need to enhance the capabilities of people to help themselves and others.

For those on the Old Left, the emergence of this fourth discourse has not gone unnoticed. There has been widespread recognition of the calls by Giddens (1998, 2000, 2002) to harness what he variously refers to as civil society or self-help. It has also been recognised that New Labour policy discourses, despite their focus upon employment-focused solutions, are starting to pay greater attention to cultivating work beyond employment. The Old Left has greeted the emergence of this fourth discourse with some cynicism if not outright opposition (e.g. Eisenschitz 1997). Although many reasons exist for their rejecting such an approach (for a review, see Giddens 2002), a principal fear is that it will lead to the creation of a 'dual society'. The perception is that while the majority will fend for themselves through paid employment, those left over and unable to insert themselves into the formal labour market will find themselves increasingly consigned to relying on alternative coping mechanisms as the welfare state is rolled back and citizens' rights diminished (e.g. Amin *et al.* 2002).

We agree with such fears concerning the possible development of a dual society, especially if this valorisation and fostering of work beyond employment is reserved for those excluded from the mainstream formal labour market. Indeed, examining the increasing number of government policy documents on harnessing such activity as a solution to poverty (e.g. DETR 1998, DfEE 1999, DSS 1998, Home Office 1999, Social Exclusion Unit 1998, 2000) suggests that the Old Left has cause for concern. These documents lean heavily towards introducing such coping mechanisms only amongst the jobless and those living in deprived neighbourhoods rather than to society as a whole. There is thus every reason for the Old Left to express grave concerns about the advent of a dual society. It is not hard to conclude that such mechanisms will be a 'second class' form of welfare provision for the poor as the welfare state realigns itself to reproduce only those most useful to the formal economy.

However, this does not have to be the case. As this book will display, the fact that one variant of the discourse on valuing and harnessing people's capacities to engage in work beyond employment may produce a dual society is no excuse for throwing out the baby with the bath water. To draw an analogy, just because job creation under neo-liberalism may result in the growth of the 'working poor', this is no reason to reject the notion of employment creation *per se*. The fact that fostering work beyond employment can lead to a dual society if it is only promoted amongst the poor and socially excluded is thus no reason to dismiss this development path and place one's hopes in a restoration of a full-employment/comprehensive welfare state scenario. To do so is grounded in dangerous 'golden age' thinking that seeks the return of an era that, as we will show, never actually existed.

In order to situate this book, therefore, in this introduction we commence by briefly reviewing the principal streams of thought that have propagated this alternative third way approach that believes it necessary to value and develop forms of activity beyond employment. This will offer some under-standings of both the sources of such ideas and why analysts adopt such a stance. We then move on to consider what is meant by 'work beyond employment' and to explain why and how this book intends to articulate and develop this alternative third way approach towards tackling poverty.

SITUATING THE BOOK: THIRD WAY DISCOURSES THAT SEEK TO VALUE AND DEVELOP WORK BEYOND EMPLOYMENT

As already indicated, some of the most prominent UK social policy com-mentators are beginning to sketch out an alternative third way to the employment-centred social integrationist discourse that dominates New Labour thinking on poverty (e.g. Giddens 1998, 2000, 2002, Jordan 1998, Jordan and Jordan 2000, Levitas 1998, Lister 2000). Until now, however, little attempt has been made either to more fully develop this alternative third way or to ground it in previous bodies of thought. Here, therefore, we commence our articulation and development of this approach by briefly reviewing two bodies of thought that have adopted this discourse. First, we introduce the radical European social democratic tradition that has called not only for greater recognition and valuing of work beyond employment but also for its promotion as a means of developing an alternative and/or comple-ment to profit-motivated market exchange. Second, we review a radical ecology approach that advocates the fostering of such work in order to bring about sustainable economic development.[1] In outlining these two bodies of thought, our intention is not to intimate that those presently developing such ideas in the UK (or elsewhere) somehow adhere to them. Rather, it is solely to highlight that they are not alone in their desire to decentre formal employment from its pivotal position when forging new social models for the twenty-first century.

Radical European social democratic thought

Bubbling under the surface of mainstream European social democratic discourse, there has been a strong current of radical thought that advocates the development of work beyond employment as an alternative and/or complement to employment (e.g. Aznar 1981, Beck 2000, Delors 1979, Gorz 1999, Lalonde and Simmonet 1978, Rifkin 1995, Sachs 1984). For these radical European social democrats, there are at least three reasons for seeking to relativise the place of formal employment in the lives of all citizens and to develop alternatives or complements to the formal labour market. First, they maintain that employment is structured in a manner that is stultifying, alienating and leaves those in jobs with little energy to compensate for this outside their hours of employment. Second, employment is seen as having become structured in such a manner because society has lost its way. What were once means to an end (e.g. economic growth, having a job) have now become ends in themselves. Third and finally, they believe that the present-day crises surrounding employment (e.g. achieving full-employment, increasing flexibilisation) are thus not problems to be overcome but opportunities to rethink the current organisation of work.

For these radicals, the way forward is not more fully to insert people precisely into the profit-motivated exchange relations of employment that they view as causing so many problems (i.e. employment-focused social integrationism). Instead, they assign activity beyond employment a crucial role in the future of work and welfare. However, they are not only, or indeed, necessarily, concerned with such activity as it exists today, but with the possible emergence or reinforcement of a category of work that one could call 'autonomous' in the future. Although some radicals view the green shoots of such autonomous work in the present-day informal sphere (e.g. Jordan 1998, Rifkin 1995), others view autonomous work as a conceptual, as opposed to concrete, phenomenon (e.g. Gorz 1999, Sachs 1984). For these latter analysts, autonomous work is not currently a tangible, empirically observable category of activity. Instead, it is a new form of socially useful work that should be created where the producer has control over the work and in which creativity and conviviality will be the driving forces. This form of work will have a purpose for the person performing it other than earning a wage (for a summary, see Windebank 1991).

What unites all of these radical European social democrats, nevertheless, is their wish to put an end to, or at least considerably reduce, the domination of 'heteronomous' work over our lives. Heteronomous work is understood as those productive activities over which individuals have little or no control and is characterised by formal employment. Some radicals with leanings towards the Old Left argue that this can be achieved by the state taking control of heteronomous production and designating the boundaries between the autonomous and heteronomous spheres of life. Increasingly, however, many radicals are instead seeking changes to work patterns through piecemeal civil society initiatives that will nibble away at the domains of the formal market and/or public provision. In adopting this approach to change, these radical European social democrats overlap considerably with a second separate but interrelated stream of thought that is also seeking to value and cultivate work beyond employment.

Communitarian ecocentrism

The desire to challenge the domination of formal employment in society is not only advocated by radical European social democrats. There is also a large body of green political thought that seeks to recognise and foster work beyond employment in order to achieve sustainable economic development (e.g. Dobson 1993, Ekins and Max-Neef 1992, Fodor 1999, Goldsmith *et al.* 1995, Henderson 1999, Hoogendijk 1993, Mander and Goldsmith 1996, McBurney 1990, Robertson 1991, Roseland 1998, Trainer 1996, Warburton 1998, Wright 1997).

Grounded in a broader green political ideology, these analysts view socialism as a spent political force as displayed by both the changes in the political economy of Central and Eastern Europe and the parallel shifts in the politics of First World 'advanced' economies. For them, this is not something to be mourned. The old quarrels between neo-liberals and socialists were simply over the best way of boosting productivism and realising greater materialism for the majority of people. The advent of New Labour's third way grounded in employment-centred social integrationism continues in the same vein, merely introducing a further alternative to free market capitalism now that socialism is dead. For these greens, the differences between all of these approaches are differences that make no difference. The pursuit of greater materialism and enhanced productivism displays how what were originally means to an end have become ends in themselves in these approaches (e.g. Capra and Spretnak 1985, Dobson 1993, Mander and Goldsmith 1996, Robertson 1991). For these radical ecologists, there is a need to recapture the ends. To do this, it is argued that we need to reconsider first, the relationship between people and nature and second, and flowing from this, the direction of society (e.g. Devall 1990, Eckersley 1992, Goodin 1992).

Indeed, it is precisely due to their arguments concerning how the relationship between people and nature needs to be rethought, and how society needs to be restructured, that we refer to these analysts as 'communitarian ecocentrists' (see O'Riordan 1996) in this volume. Rather than protect natural ecosystems simply for the pleasure of people (i.e. anthropocentrism), they adopt an ecocentric approach viewing nature as having biotic rights that require no justification in human terms (e.g. Capra and Spretnak 1985, Devall 1990, Devall and Sessions 1985, Naess 1986, 1989, Skolimowski 1981). Flowing from this ecocentrism, their argument is that ecologically sustainable development can occur only if we pursue a more small-scale decentralised way of life based upon greater self-reliance (e.g. Douthwaite 1996, Ekins and Max-Neef 1992, Gass 1996, Goldsmith *et al.* 1995, Henderson 1999, Lipietz 1995, Mander and Goldsmith 1996, McBurney 1990, Morehouse 1997, Robertson 1985, Roseland 1998, Trainer 1996). To achieve this, the now established concept of 'thinking globally and acting locally' is the key. They believe that global problems such as the destruction of nature can be overcome only by acting in a local manner (e.g. Mander and Goldsmith 1996). Rather than pursue the end of economic growth through outward-looking development policies, their objective is instead to develop more 'inward-looking' approaches focused upon meeting local basic needs through the pursuit of self-reliance (e.g. Ekins and Max-Neef 1992,

Robertson 1985, Morehouse 1997). The development of work beyond employment thus resonates with their desire for more localised, self-reliant and sustainable economic development (e.g. Henderson 1999, Mander and Goldsmith 1996, Warburton 1998).

WHAT IS MEANT BY WORK BEYOND EMPLOYMENT?

New Labour's employment-centred social integrationist discourse is thus not the only way of paving a third way beyond the Old Left and the New Right. A fourth discourse on resolving poverty is evident, which not only has foundations in both radical European social democratic and green political thought but also is starting to be articulated amongst prominent UK social policy commentators. If poverty is to be alleviated and a more inclusive society developed, the assertion is that integration into employment is insufficient. Instead, society must focus on fostering the capabilities of people to participate in work beyond employment so as to harness their capacities to help themselves and others. In order to further elucidate and develop this discourse, we start by unpacking what is meant by work beyond employment.

For all of these analysts, 'work beyond employment' refers to economic activities that lie beyond the nexus of formal market-orientated production and/or exchange grounded in the social relations of employment. As the term implies, therefore, 'work beyond employment' is defined by what it is not. It is the other (and 'othered') work that is not employment. Given that this is what it is, no other definition is possible. Indeed, the fact that such work can only be defined in relation to what it is not is crucial. It accurately reflects that in our society in which profit-motivated market-like exchange relations are seen as hegemonic, all work that is leftover is cast into a residual catch-all umbrella category.

In order to help overcome such a market-centred classification of work, it is first necessary to adopt some more positive terms to describe productive activities that are not formal employment. Despite the growing prevalence of terms like 'civil labour' and 'self-help' (e.g. Beck 2000, Williams and Windebank, 2001b), the most common terms used to refer to work beyond employment are the 'informal economy' or 'informal sector'. However, in this book, although we will assign the adjective 'informal' to work beyond employment, we do not use either of these nouns. We adopt the adjective 'informal' quite simply because it is the most popular term used to describe all forms of paid and unpaid work existing outside employment. There is no other term currently in usage that provides such instant recognition of what is being discussed. It is also perhaps one of the most accurate adjectives that could be used to portray how the social relations in this realm of economic life differ from the more 'formal' social relations in which official employment is embedded.

However, the noun 'economy', which is often attached to the adjective 'informal' is not used in this volume. It is increasingly apparent that these activities do not themselves constitute a separate economy (e.g. Harding and Jenkins 1989, Thomas 1992, Williams and Windebank 1999b). As Gershuny (1985: 129) asserts, 'the informal economy . . . is of course not a separate economy at all but an integral part of the system by which work, paid and unpaid, satisfies human needs'. To speak of work outside employment as an 'economy' implies that it enjoys a degree of autonomy from the market. As this book argues throughout, nevertheless, there is an intimate interdependent relationship between informal and formal economic activities. For this reason, any reference to such work as a separate 'economy' has been deliberately avoided.

The problem with the term 'sector', meanwhile, is that both in everyday language and in the Standard Industrial Classification (SIC) index, a sector is a sphere in which a particular type of good is produced or service offered. Informal work, nevertheless, cannot and is not defined in such a way. Such work is not constituted by a particular set of tasks or activities but cross-cuts all sectors. Informality, after all, is not an inherent property of specific activities (e.g. housework). All goods and services can be produced and distributed either formally or informally. As such, informal activity is not a sector.

In consequence, we use the adjective 'informal' in conjunction with the terms 'work', 'economic activity' or 'modes of production' when referring to all of the work beyond employment. It is important, however, especially if the messiness and complexity of 'the economic' in contemporary society is to be more fully understood, that the heterogeneous economic activities included under this catch-all umbrella are differentiated. To do this, three slightly more coherent categories based on the social relations within which such work is conducted are identified (e.g. Gregory and Windebank 2000, Leonard 1998, Pahl 1984, Renooy 1990). These are:

- 'self-provisioning', which is the unpaid work undertaken by household members for themselves and each other. This ranges from domestic labour through unpaid caring activities conducted for and by household members to do-it-yourself home improvements;
- 'unpaid community work', which is work provided on an unpaid basis by the extended family, social or neighbourhood networks and more formal voluntary and community groups; and
- 'paid informal exchange' where legal goods and services are exchanged for money and gifts, which are unregistered by, or hidden from, the state for tax, social security or labour law purposes.

Although this differentiation of informal work into these three principal categories is adopted throughout this book, we recognise that there is still great diversity within each category. Wherever it is revealing to do so therefore, we divide each of these three forms of informal work into a number of further subcategories.

Self-provisioning, or what Polanyi (1944) called 'householding', can include a diverse range of activities ranging from routine tasks such as washing-up to more creative and non-routine tasks such as carpentry. These are often conducted under very different social relations and the motives underpinning participation often lie in stark contrast to each other. Here, in consequence, and in response to much of the literature that has called for a distinction to be drawn for analytical purposes between routine domestic work, do-it-yourself (DIY) activity and care work (e.g. Gardiner 1997, Gregory and Windebank 2000, Walby 1997), we divide such activities into these three subcategories wherever appropriate.

Unpaid community work, similarly, will be subdivided wherever it is pertinent to do so for analytical purposes into three types: kinship work, friendship/neighbourly work, and voluntary work. Although there is a tendency in some analyses to further subdivide voluntary work into numerous categories (e.g. Burns and Taylor 1998, Home Office 1999), this is not considered to be necessary for our purposes here. As will be revealed, voluntary work is used to such a minor extent by households to mitigate their poverty and social exclusion as to be hardly relevant and as such, does not warrant further subdivision. Micro-level reciprocal exchange between kin, neighbours and friends in contrast is much more heavily relied upon and given the variations in the attitudes towards, and motives underpinning participation in such activity, it is divided into unpaid kinship exchange and unpaid exchange between friends/neighbours.

The third and final form of informal work, paid informal exchange, refers to the paid production and sale of goods and services that are unregistered by, or hidden from the state, for tax, social security and/or labour law purposes but which are legal in all other respects (e.g. European Commission 1998a, Feige 1990, Portes 1994, Thomas 1992, Williams and Windebank 1998a). As such, it covers only activities that are illegal because of their non-declaration to the state for tax, social security and/or labour law purposes. It excludes activities in which the good and/or service itself are illegal (e.g. drug trafficking).

This activity is often associated with profit-motivated monetary exchange and seen as an exploitative form of employment conducted by marginalised populations as an economic survival strategy. In this book, however, we show that this is not always the case. It is not correct simply to align this form of economic activity to market-like work. Instead, we argue that it bridges the market/non-market divide depending on the types of the social relations in which activities are embedded and the motives underpinning them. We distinguish, therefore, two forms of paid informal exchange. First, there is 'non-market'-orientated paid informal work, where people engage in paid exchange mostly for friends, relatives and neighbours. Second, there is 'market-like' work conducted on a paid informal basis that occurs where an employee conducts work for a formal or informal business under social relations akin to formal employment for profit-motivated rationales.

With this discussion of the scope of what is being discussed in hand, attention now turns towards outlining how this book intends to articulate and develop this alternative third way approach.

STRUCTURE OF THE BOOK

The intention of this book, to repeat, is to display that third way discourses beyond the Old Left (the 'first way') and the New Right (the 'second way') are not limited to New Labour's approach of social integration through paid employment. Our aim is to articulate and further develop an alternative third way. This asserts that alleviating poverty and building a more inclusive society is more than merely a matter of providing the poor with formal jobs. If the capabilities of people to help themselves and others are to be developed, we also need to give them access to participation in informal work.

Until now, this approach has been only partially developed. Consequently, this book attempts to elaborate this emerging discourse in three steps. Part I outlines the rationales for adopting this alternative third way approach. Part II then develops a methodology for studying poverty that focuses upon capabilities and takes fully into account the role of informal work in fostering capabilities. Finally, Part III sets out a new policy agenda for combating poverty based on harnessing this sphere in order to cultivate the capacities of people to help themselves and others.

Rationales for an alternative third way

Despite, or perhaps because of, the recent interest shown in work beyond employment by those attempting to transcend old-style social democratic thought (Beck 2000, Giddens 1998, 2000, 2002), the Old Left have been amongst the most vocal critics of those espousing this alternative third way. For them, the attempt to foster the capacities of people to help themselves and others is advocated for at least two purposes. On the one hand, it is seen as a ploy to reduce welfare costs in an era of increased global competition. On the other hand, it is viewed as part of an ideological swing in welfare policy from a rights-based system to one founded on duties or responsibilities (see Jordan 1998). As Eisenschitz (1997: 160) puts it, 'Self-help legitimates the disengagement of the state from welfare pointing towards the informal economy as a replacement.' For these old-style social democrats, a belief is retained in a return to full-employment and/or comprehensive welfare provision by the state.

In Part I of this book, however, we argue that these critics first need to consider the feasibility of returning to full-employment, as well as what is being advocated by those who seek to harness work beyond employment, before rejecting this approach. To do this, Chapter 1 outlines the employment problem confronting advanced economies, Chapter 2 the stalled formalisation of the advanced economies and the growing importance of work beyond employment and Chapter 3 the various policy options towards informal work and their implications. This will reveal that opposing the development of informal work, fearing that it will lead to a demise of public sector provision and social rights, and intensified inequalities, displays myopia towards the structural problems confronting advanced economies as well as a limited understanding of the diverse approaches towards cultivating such work.

Examining poverty in contemporary England

Few studies have so far attempted to study poverty from a methodological position that focuses upon capabilities and fully takes into account the role of the informal sphere in fostering these. In Part II, therefore, we first develop a methodology that does this and then display the outcomes of so doing. Chapter 4 thus begins by developing a method to measure household poverty in terms of coping capabilities rather than the more usual measures such as income. The second part of this chapter then reports the results of adopting such a research methodology. This reveals the significant socio-spatial variations in the coping capabilities of households. In Chapter 5, we then explore the means by which households meet their needs and desires by examining the coping practices that they adopt. This reveals not only how those unable to participate in the formal sphere are also unable to partake in the informal sphere but also how the meanings attached to participation significantly vary socio-spatially. For poverty to be alleviated and a more inclusive society developed, therefore, policy interventions are required to integrate people into not only the formal but also the informal sphere. Given that policy interventions have so far largely concentrated only on how people can be inserted into employment, Chapter 6 identifies the nature of the policy interventions required to tackle the barriers that prevent households from participating in informal work.

Tackling poverty: policy initiatives for an alternative third way

Having highlighted in Part II how the socio-spatial disparities produced by employment are consolidated, rather than reduced, when informal work is incorporated into analyses of poverty, Part III starts to set out a new agenda for tackling poverty and building a more inclusive society. In the employment-centred social integrationist discourses of New Labour, the intention has been to alleviate poverty by inserting the unemployed into the formal labour market with the aim of achieving full-employment. In Chapter 7, however, it is argued that to put all of one's eggs into this basket is unrealistic. Given the importance of informal work in the coping practices of households and the apparent macroeconomic shift towards such work, our argument is that there is a need to replace the goal of full-employment with one of 'full-engagement'. By a full-engagement society, we mean one in which there is sufficient work (both employment and informal economic activity) and income so as to give citizens the capacity to satisfy both their basic material needs and creative potential.

Rather than focus upon the well-trodden path of developing policies to enhance the capabilities of people to participate in the formal labour market, Chapters 8 and 9 of this book instead investigate the so far under-explored policy area of improving the capabilities of people to participate in informal economic activity. In Chapter 8, in consequence, we show how the role of bottom-up initiatives radically changes once the goal shifts from full-employment to full-engagement. Rather than attempt to harness the more formal third sector organisations as springboards into employment, full-engagement refocuses attention on identifying bottom-up initiatives that can harness one-to-one acts of reciprocity. To evaluate possible 'fourth sector' initiatives that might fulfil this objective, this chapter reviews local exchange and trading schemes (LETS), time banks, employee mutuals and mutual aid contracts.

Until now, these have been viewed in policy circles more as a bridge into employment than as a tool for facilitating informal economic activity (e.g. DfEE 1999, Social Exclusion Unit 2000). However, we reveal that although such initiatives can provide a springboard into the formal labour market, they are much more effective at enabling people to help themselves and others. If full-employment is the goal, therefore, such initiatives have only a minor role to play in tackling poverty and are of marginal significance. If, however, the goal is to bolster the capabilities of people to help themselves and others by providing alternative means of livelihood in the form of informal work, then these initiatives have a more central place on the policy agenda. As will be displayed, nevertheless, such bottom-up initiatives although necessary, are insufficient alone to tackle poverty and social exclusion.

Chapter 9 thus considers a variety of top-down initiatives that might have a larger impact on enabling people to help themselves and others. In this chapter, we consider initiatives that shift the 'worker citizen' model from one in which employment is the sole source of security, esteem and identity to one where the working citizen is integrated not only through employment but also through other forms of active citizenship. To do this, consideration is given to several top-down initiatives including citizens' income and various forms of 'active community service' that could record, store and reward participation in such activity.

CONCLUSIONS

In sum, our aim in this book is to articulate and develop an alternative third way approach towards tackling poverty and social exclusion that, although only briefly sketched up until now, is gaining greater recognition as a potential way forward. This rejects the obsession of New Labour with inserting people into employment as a means of tackling their poverty (e.g. Beck 2000, Giddens 1998, 2000, Jordan 1998, Levitas 1998, Lister 2000). Instead, its basic premise is that to alleviate poverty and build a more inclusive society, the capabilities of people to participate in alternative coping practices in the form of informal work need to be developed. Therefore, just as it is widely accepted that intervention in the formal labour market is required in order to alleviate poverty, the argument of this book is that informal work also needs the aid of initiatives. Why this has to occur and how it can be achieved is the subject of this book.

Note

1   There is also a long-standing feminist tradition that calls for the recognition and valuing of work beyond employment (see Chapter 2). This tradition, however, and unlike the two bodies of thought considered here, does not lead its adherents to seek its further development. Instead, most 'old style' feminists shy away from promoting such work because they believe that women's insertion into employment is the key to their liberation (see Gregory and Windebank 2000). In this regard at least, the ideas of 'old style' feminism resonate much more strongly with New Labour's view that insertion into formal employment is the route to a more inclusive society rather than with the alternative third way discourse articulated and developed in this book.

Part I

Rationales for a third way approach

1 The problem of full-employment

Throughout the advanced economies, the widespread consensus is that employment is the best route out of poverty. Not only are the approaches of both the Old Left and New Right grounded in such a belief, but so too is the employment-focused third way approach of New Labour. In this book, however, our intention is to begin to explain why an alternative third way discourse has started to emerge that rejects an employment-centred approach to poverty alleviation (e.g. Beck 2000, Jordan 1998, Levitas 1998, Lister 2000). To do this, we evaluate critically first, the feasibility of achieving full-employment and second, whether this approach towards poverty alleviation is desirable.

To commence, therefore, this chapter addresses whether 'full-employment' is a feasible objective. This will show that, despite all of the efforts to create formal jobs, there remains a large gap between the current employment situation and a full-employment scenario. This jobs gap, moreover, is not narrowing over time. Indeed, there appears to be little hope either in the near future or even beyond, that a state of full-employment will be achieved. Following this, attention turns to the desirability of pursuing full-employment. On the one hand, we analyse how such an approach does not appear to result in a reduction in poverty by examining the prevalence of the 'working poor'. On the other hand, the changing attitudes towards employment in contemporary society are examined. This uncovers that although it is often taken for granted in policy-making that people want a formal job, it is actually the income from the job rather than the job itself that people desire. Indeed, the trend in the advanced economies appears to be one of a centring of income but a de-centring of employment in many people's lives.

Such findings, of course, raise many issues that will need to be addressed later in this book. If we cannot achieve full-employment, if many people entering the formal labour market are continuing to witness poverty and there is a de-centring of the importance of employment in many people's lives, then should full-employment continue to be our prime social objective? Indeed, is it logical that a third way approach, which supposedly seeks to move beyond capitalism and socialism, should have as its focus the mass insertion of people into precisely those profit-motivated exchange relations that it is seeking to transcend? Are there any alternative routes out of poverty besides entry into formal employment? If so, what are these alternatives and is it possible to cultivate them? And what are the implications of doing so? Before answering such questions, however, it is first necessary to outline the problem of full-employment in our advanced economies. Indeed, it is only once this problem has been firmly established that the importance of seeking alternative routes out of poverty can be fully understood.

THE FEASIBILITY OF FULL-EMPLOYMENT

In recent years, many writing from positions closely linked to European radical social democratic thought and/or communitarian ecocentrism (see the introduction) have questioned whether full-employment is an achievable goal (e.g. Beck 2000, Bridges 1995, Giddens 1998, Gorz 1999, Rifkin 1995, Williams and Windebank 1999a, 1999b). Bridges (1995), for example, argues that those looking back at the end of the twenty-first century will view the current preoccupation of governments with inserting people into jobs as akin to finding deckchairs for everybody on the *Titanic*. Beck (2000) similarly contends that it is futile to hark back to the supposedly golden age of full-employment. Instead, he suggests that we should use the demise of a full-employment society as an opportunity to develop new ideas and models for work rather than look to previous 'golden ages' for our inspiration of the way forward. Gorz (1999: 58), in a cutting criticism of those seeking a return to a full-employment society, goes further:

> Those who continue to see work [employment]-based society as the only possible society and who can imagine no other future than the return of the past . . . do everyone the worst service imaginable by persuading us that there is no possible future, sociality, life or self-fulfilment outside employment, by persuading us that the choice is between a job and oblivion, between inclusion through employment and exclusion, between 'identity-giving socialization through work' and collapse into the 'despair' of non-being. They persuade us it is right, normal, essential that 'each of us should urgently desire' what in actual fact no longer exists and will never again lie within everyone's grasp: namely 'paid work in a permanent job', as the 'means of access to both social and personal identity', as 'a unique opportunity to define oneself and give meaning to one's life'.

Given the recent successes in job creation throughout the advanced economies and the significant reduction in official unemployment rates, some might question the validity of such a pessimistic view. To explain the reasons for such doubts concerning the feasibility of full-employment, therefore, this section investigates first, the size of the 'jobs gap' in the advanced economies between current employment participation rates and a full-employment scenario and second, whether this gap is declining or increasing over time. This will display that for all of the current hyperbole in advanced economies such as the US and UK about full-employment being within our grasp, advanced economies are far from achieving this goal.

Employment participation rates

Examining employment participation rates in the advanced economies, the principal historical lesson is that full-employment was achieved for at most thirty years or so following World War II in a handful of advanced economies (Pahl 1984). Even here, however, this was only full-employment for men, not women (Gregory and Windebank 2000). As Beveridge (1944: 18) put it, full-employment is a state in which there are 'more vacant jobs than unemployed *men*' and where there are jobs 'at fair wages, of such a kind, and so located that the unemployed *men* can reasonably be expected to take them' (our emphasis). Although this language was of course based on the widely accepted sexist prose used at the time, he was in effect quite correct in his statement when referring to the full-employment of men alone. Any talk of returning to a 'golden age' of full-employment is illogical if by that is meant an era of full-employment for both men and women. Such an era has never existed so to seek its return is not possible.

If seeking a return to full-employment for the entire adult population is illogical since it never existed, it is perhaps also the case that it is not possible. Take, for example, the UK. By May 2000, according to the *Labour Force Survey*, the number of people in jobs had reached a record level of 27.8 million and the unemployment rate, based on the claimant count, was down to a mere 3.9 per cent (5.8 per cent using the wider *Labour Force Survey* measure). Superficially, this appears close to what is conventionally meant by 'full' employment when the unemployment rate is 2–3 per cent to allow for the churning of people between jobs. However, despite the official unemployment rate being at its lowest level for decades and UK employment participation rates being amongst the highest in Europe (second only to Denmark) and at a record level, only 74.4 per cent of the working-age population were in jobs. Over one in four working-age people (25.6 per cent) did not have a job. In other words, to achieve full participation in employment of the UK working-age population, one additional job would be needed for every three in existence (a 33 per cent rise), the equivalent of some 9 million additional jobs.

Indeed, compared with the European Union as a whole, this UK data on the size of the 'jobs gap' that needs to be bridged is quite hopeful. In 1999, just 147 million of the 375 million inhabitants of the EU were in employment (40 per cent of the total population). Some 60 per cent of the population were thus being supported by the remaining 40 per cent and there is widespread agreement that this is due to worsen as the 'baby-boom' generation reach retirement age. Amongst the population of working age, meanwhile, the employment participation rate in the EU is just 61 per cent (European Commission 2000b). Nearly two in five (39 per cent) of the working-age population in the EU are thus without a job. For full participation to be achieved in the EU, two jobs are thus needed for every three that currently exist. Put another way, a 66 per cent increase in the number of jobs is required. Some EU nations, nevertheless, have a smaller bridge to cross to reach this supposed nirvana than others. In Denmark, the country with the highest employment participation rate in the EU (75.3 per cent), due in no small part of a very high participation rate of women in employment, a mere 33 per cent rise in the number of jobs would suffice. In Spain, the country with the lowest employment participation rate (49.7 per cent), however, the number of jobs would need to double (European Commission 2000b).

It might be asserted that presenting employment statistics in this manner is a distortion. For example, it ignores the considerable number of people who now engage in higher education so as to feed the new 'knowledge economy', the army of mothers at home supported by an employed spouse, and so forth. It might also be asserted that it is far better to investigate the trends over time so that the progress being made towards full-employment can be seen, rather than just provide a snapshot of a particular moment. Here, therefore, we examine both issues in turn, starting with the latter.

Is it the case that the inexorable long-term trend is towards full-employment? Examining the official data, this is not the case. As Table 1.1 displays, during the last forty years of the twentieth century few EU nations managed to make any progress in closing the 'jobs gap'. Indeed, just three nations made any progress at all. Denmark managed to raise employment participation rates from 74 per cent to 76 per cent, the Netherlands increased them from 64 per cent to 71 per cent, and Portugal, starting from the low base-level of 58 per cent participation, managed to raise it to 67 per cent. These, however, are the exceptions. The majority of nations went backwards over this forty year time period. The 'jobs gap' widened. Some falls were quite dramatic. Employment participation rates in Finland slid from 77 per cent in 1960 to 67 per cent in 1999. In France they fell from 70 per cent to 60 per cent, in Greece from 66 per cent to 55 per cent and in Italy from 65 per cent to 53 per cent. The notion that there is a long-term trend towards full-employment, therefore, must be treated with considerable caution. It is not borne out by the evidence.

If EU nations are burdened by a 'jobs gap' that in most cases has widened over the past forty years, is it nevertheless the case that this is not mirrored in other First World nations, especially those where the neo-liberal project has taken a firmer hold? To answer this, let us start with the US. After all, this is the major competing trading bloc that provides the 'baseline' against which the EU measures its progress on employment participation rates (see European Commission 2001a). In the US, there is little doubt that the employment participation rate is higher than in the EU. However, even here

Table 1.1 Labour force participation rates, 1960, 1973 and 1999

Country	Total participation rate			Growth (+) or decline (−) 1960–1999
	1960	1973	1999	
Finland	77	71	67	−
Sweden	75	75	71	−
Denmark	74	75a	76	+
UK	72	73	71	−
Austria	71	69	68	−
France	70	68	60	−
Germany	70	69	65	−
Ireland	68	64	63	−
Greece	66	57	55	−
Italy	65	58	53	−
Netherlands	64	61	71	+
Spain	61	60	52	−
Belgium	60	61	59	−
Portugal	58	64	67	+

Source: ILO (1997: Table 2.2) and European Commission (2001a: Annex 1).

it is only 73 per cent, meaning that one job needs to be created for every three that currently exist if full participation is to be achieved (European Commission 2000b). Put another way, this requires a 37 per cent increase in the number of jobs in the US economy. In Japan, similarly, the employment participation rate is 70 per cent, necessitating a 42.8 per cent growth in the number of jobs to achieve full participation (European Commission 2000b).

Across the advanced economies, there is thus a wide gap between current employment levels and a situation of full participation. Nor is the trend narrowing over time. Between 1960 and 1995, just 13 of the 22 advanced economies improved their employment participation rates. The outcome was that by 1995, only 9 of the 22 advanced economies had managed to achieve participation rates in employment of over 75 per cent of the working-age population and none over 83 per cent (ILO 1997). Advanced economies, therefore, are far from a steady state of full-employment and many are moving ever further away from such a situation.

So too are many deprived regions and localities. In the EU, for example, the 25 EU regions with the lowest unemployment rates were much the same in the late 1990s as in the late 1980s and their unemployment rates remained steady at around 4 per cent. The 25 regions with the highest unemployment rates were again much the same in the late 1990s as in the late 1980s, but their unemployment rates had increased by, on average, 20 per cent (European Commission 1999b). Over time, in consequence, the regions with the highest and lowest unemployment rates have further polarised. Those with the widest jobs gap in the late 1980s had fallen further behind by the late 1990s. This polarisation is similarly evident within the UK. Regions with the fastest growing employment participation rates are those which already had high employment participation rates, namely the South East, South West and East Anglia. The result is a regional divergence in the size of the jobs gap (Dunford 1997).

The same applies on a micro-spatial level. In the UK, wards with relatively high employment participation rates are compounding their advantages over time, while those starting with relatively low participation rates are falling ever further behind them (Dorling and Woodward 1996, Dunford 1997, Green and Owen 1998). As Green and Owen (1998: viii) put it, the

> largest increases in unemployment, and especially in inactivity and non-employment, [are] in the neighbourhoods – particularly those in inner-city areas and in concentrations of public sector housing – where the initial incidence was highest.

In consequence, despite concerted efforts to boost employment participation rates in deprived regions and localities, they are falling ever further behind. The net result is that the non-employed are becoming more spatially concentrated. In 1981, about 43 per cent of the British working-age population lived in wards where the proportion not employed varied by less than 10 per cent from the national average. This proportion had fallen to 30 per cent by 1991, indicating a divergence between areas (Dorling and Woodward 1996). People have become more likely either to live in a ward where many other adults of working age do not have a job, or to live in a ward where the large majority of adults of working age are employed.

Conventional unemployment measures, moreover, underplay not only the true extent of inactivity, non-employment and under-employment, but also the acuteness of the spatial disparities. Green and Owen (1998) show that the greater the degree of labour market disadvantage in an area, the smaller is the proportion of the inactive and non-employed who are included within conventional definitions of unemployment. As such, inactivity, unemployment and non-employment are far higher, especially in deprived neighbourhoods, than is suggested by headline unemployment statistics. So too is the gap between deprived and affluent localities wider than that suggested by data on the registered unemployed.

Examining employment participation rates, a large jobs gap can be thus identified between the current employment situation and a full-employment scenario and this jobs gap is growing in many advanced economies. Despite the strenuous efforts of public policy to bolster employment in deprived regions and localities, moreover, such areas appear to be falling ever further behind.

Such a portrait of employment participation rates paints a gloomy picture concerning the feasibility of full-employment. Although we could end our analysis at this point, it is nonetheless necessary to point out one further important trend that brings little comfort to anybody wanting to rely on insertion into employment as the principal tool for resolving poverty. So far, we have considered only whether or not people have a job. We have not considered the types of employment they undertake. When analysed, however, the gloom concerning the feasibility of full-employment turns into despair. The above statistics on employment participation rates omit to consider that, over time, an increasing proportion of those counted as 'in employment' are in part-time rather than full-time jobs (e.g. European Commission 2000b, Nicaise 1996, Thomas and Smith 1995, Townsend 1997). In the EU, for example, the share of part-time employment increased from 14 per cent of all employment in 1990 to 17 per cent in 1998 (European Commission 2000b). This historical shift in employment contracts from full- to part-time is prevalent across the advanced economies. The outcome is that using employment participation rates alone to analyse whether or not there is a move towards full-employment masks the extent to which underemployment is rising. The employment problem in the advanced economies is one of achieving not just full-employment but also full-time employment.

Is it the case that this interpretation is too pessimistic? Are we, for example, ignoring that a large number of the non-employed are actually engaged in higher education so as to feed the new 'knowledge economy', are mothers at home supported by an employed spouse and so forth?

Who are the non-employed?

Using data from the mid-1990s, Hirsch (1999) identifies the nature of the 10 million working-age jobless in the UK. He finds that about two million are in full-time further and higher education, three million are women who are married or cohabiting and not employed and five million are adults living in households without anybody in a job (which also contain just over two million children). Of these, slightly under two million are actively looking for jobs, three million are disabled people not working and not looking for work and around one million are non-employed lone parents (see also Dorsett 2001).

In consequence, half of the non-employed are not wives of employed spouses and/or those in education. These five million people are living in jobless working-age households. Indeed, this is also the case beyond Britain. Throughout the advanced economies, one of the most significant contemporary labour market trends is the demise of the single-earner household and the polarisation of society into no-earner and multiple-earner households. Examining the distribution of employment across prime age (20–59) households for 13 OECD countries between 1983 and 1994, Table 1.2 reveals the extent of this process. The polarisation of society into households where none are in employment and households where all are in employment is prevalent across all OECD nations (with the exception of Canada). However, in only

Table 1.2 The polarisation of employment between households, OECD nations, 1983–94

% of all households	Jobless households			Mixed employment status households			Households where all are in employment		
	1983	1990	1994	1983	1990	1994	1983	1990	1994
UK	16.0	14.3	18.9	30.1	22.0	18.6	53.9	63.7	62.1
US	13.1	10.0	11.5[d]	32.3	24.9	24.9	54.6	65.1	63.6
Germany	15.0[a]	12.8	15.5	32.5	27.7	25.6	52.5	59.5	58.9
Netherlands	20.6[b]	17.2	17.2	39.1	31.9	27.0	40.3	50.9	55.7
France	12.5	14.4	16.5	30.6	28.3	27.9	56.9	57.4	55.7
Belgium	16.4	18.0	19.6	41.8	33.7	28.8	41.8	48.3	51.6
Australia	11.9[c]	14.5[e]	—	32.6	28.3	—	55.8	57.2	—
Portugal	12.7[c]	10.8	11.0	38.3	32.9	32.6	49.0	56.4	56.4
Canada	15.2	12.5	15.1	35.7	37.0	35.9	49.1	50.6	49.0
Ireland	17.2	20.0	22.3	47.3	40.8	36.9	35.5	39.3	40.9
Greece	16.0	16.9	17.6	46.3	40.1	38.9	37.7	43.0	43.5
Luxembourg	10.9	9.3	10.5	47.3	42.1	39.0	41.8	48.7	50.5
Italy	13.2	14.3	17.2	47.4	43.1	42.8	39.4	42.6	40.0
Spain	19.4[c]	15.2	10.8	54.5	51.6	48.1	26.2	33.2	31.8

Source: Gregg and Wadsworth (1996: Table 1).

Notes
a Data for 1984
b Data for 1985
c Data for 1986
d Data for 1993
e Data for 1991

eight out of the fourteen nations did the proportion of jobless households increase between 1983 and 1994. In the remaining nations, social polarisation has been a consequence of the increasing share of households where all are in employment. Although this social polarisation of households is thus a near universal trait, the rise in the proportion of jobless households is not.

Indeed, examining what has happened since 1994, Table 1.3 reveals that the proportion of the population living in jobless households has slightly declined in most nations. Nevertheless, across the EU in 2000, some 13.6 per cent of the population remained living in jobless households with no attachment to the formal labour market. A significant minority of the population is thus in households excluded from the labour market, ranging from 15.4 per cent of the total population in the UK to just 6.3 per cent in Portugal.

The existence of such a large number of no-earner households might not be so damning if the people living in them were being supported by the welfare state and not living in continuous poverty. It would show that governments were at least ameliorating the plight of these households. In the UK, however, Dickens *et al.* (2000) find that in 1996 some 70 per cent of jobless

Table 1.3 Persons living in jobless households in European Union nations (%)*

	1995	1997	2000
Belgium	16.9	16.9	15.2
Germany	14.3	15.9	15.3
Greece	11.2	10.8	10.7
Spain	16.3	15.1	11.3
France	11.4	12.0	11.2
Ireland	16.1	15.0	—
Italy	12.2	12.4	11.3
Luxembourg	10.3	10.2	9.7
Netherlands	14.3	12.6	11.5
Austria	8.0	9.0	9.9
Portugal	8.1	8.4	6.3
United Kingdom	17.7	16.8	15.4
EU-15	15.2	15.0	13.6

Source: Eurostat Labour Force Survey (cited in European Commission 2001b: 14)

*No data for Denmark, Finland and Sweden

households had less than half the mean household income whilst the figure for jobless households with children was 90 per cent. Moreover, taking the poverty line to be where annual incomes fall below 50 per cent of the median of household disposable income adjusted for household size, Oxley (1999) finds that 91 per cent of the continuously poor live in jobless households in the UK. Indeed, the average income of UK no-earner households is only 28 per cent of that in multiple-earner households (Harkness 1994).

These jobless households, moreover, are unevenly distributed across space. In the UK, Dunford (1997) and Williams and Windebank (1995a) show that jobless households are under-represented in the relatively affluent regions of Greater London, the South East, East Anglia and the East Midlands, and over-represented elsewhere. This is not to assert that the related problem of household poverty is merely a problem of deprived regions. Despite households in the lowest decile of the income distribution being under-represented in Greater London, the rest of the South East, East Anglia and the Midlands, in absolute terms, nearly one-quarter of all households in the lowest decile of the income distribution are located in these regions (Perrons 1995).

To explain the prevalence of jobless households, a good deal of emphasis has until now been placed on how the feminisation of the labour force has been articulated and mediated at the level of the household unit. It has often not been worth women continuing with their part-time jobs if their husband is on means-tested social benefits, since this would affect his entitlement to benefits (Hewitt 1994, Morris 1987). So, given the trend towards employing more women and more part-timers, the consequence of this 'disincentive effect' has been a polarisation of households. The social security regulations have militated against wives of unemployed men taking part-time employment since it would affect his entitlement to benefit. As such, it has thus only been worthwhile for the wife of a man in long-term unemployment to seek, or remain, in employment if she could earn substantially more than they could jointly claim in social security. This has been unlikely given the generally low levels of women's earnings, especially in part-time jobs. Wives of employed men, however, have been free to accept such part-time employment. The result of this type of understanding has been new policy interventions, such as the Working Families Tax Credit (WFTC) in the UK, to unshackle jobless households from this disincentive effect.

Bennett and Walker (1998), however, warn against assuming that all partners of unemployed people are jobless, or that such disincentives are the only reason that they do not work. Indeed, Elias (1997) finds that less than a fifth of the shortfall in employment among unemployed claimants' partners can be attributed to benefit disincentives. To help explain this, Morris (1995) asserts that for women, the reason most commonly stated for leaving work was pregnancy, and for non-employment, household and childcare obligations. Davies *et al.* (1992), meanwhile, argue that although the disincentive effect is relevant, the main explanation is that men who tend to experience unemployment (low skilled, unqualified and so forth) are more likely to marry women who have a low level of attachment to the labour market.

The important point about all of these explanations is that they attempt to identify the reasons for the rise in jobless households so that measures can be put in place for reinserting them into formal employment. The issue that all fail to consider is whether there are actually a sufficient number of jobs in society for everybody to be employed. Our argument is that although it is important to redistribute employment both spatially and between households, this needs to be planned in a context that recognises the chasm between current employment levels and a full-employment scenario. In other words, there needs to be recognition that one is engaged in the allocation of lifebelts in a situation where there are too few to go around all of the passengers.

Indeed, it is for precisely this reason that focusing upon employment insertion as the principal route out of poverty is short-sighted. Given the significant gap between the current employment situation and a full-employment scenario, there appears to be little hope either in the near future or even beyond that a full-employment scenario will be achieved. Most governments in advanced economies have recognised this problem even if they have not made it explicit to their electorate. Why else, for example, did the EU member states at the Lisbon European Council in March 2000 decide to redefine what is meant by full-employment? At this meeting, although the overarching objective of full-employment was maintained and reiterated, it became redefined to be:

> To realise Europe's full employment potential by working towards raising the employment rate to as close as possible to 70% by 2010.
>
> (European Commission 2000b: 15)

What no government has so far explained, however, is what is to be done with this 30 per cent of the EU population of working age who are not envisaged as potential employees? How are these 75 million people of working age to make a living? This is a crucial issue.

If policy-makers leave the goal of fuller-employment in place and seek separate solutions for this large segment of the population excluded from the labour market, then the fallout is likely to be a 'dual society'. The 70 per cent will continue to find their salvation through the formal labour market while the 30 per cent excluded will have to be given some alternative coping mechanism. This might be a passive welfare benefits system or it might be payment for active citizenship or a reinvigorated informal sphere through which they can pursue alternative means of livelihood. However, the impacts of a policy approach that maintains the goal of full-employment but introduces alternative work and welfare practices for the vast numbers who fall by the wayside is not for us the solution. Rather than deliberately create a 'dual society', it is perhaps far better to question the current goal of full-employment and consider whether some alternative goal can be put in its place. It is not just the jobs gap that leads us to this conclusion.

THE DESIRABILITY OF FULL-EMPLOYMENT

Besides issues surrounding the feasibility of full-employment, there are crucial questions concerning its desirability. On the one hand, this is because such an approach does not appear to be resulting in a reduction in poverty for many of those who manage to find a job. On the other hand, it is because attitudes towards employment in contemporary society appear to be changing.

The problem of the 'working poor'

Does insertion into employment represent a route out of poverty? According to the available evidence, a large proportion of those who manage to insert themselves into the formal labour market still find themselves confronted by poverty. As Beck (2000: 90) puts it, 'Work and poverty, which used to be mutually exclusive, are now combined in the shape of the *working poor* [our emphasis].'

As the European Commission (2001b: 99) report, employment does not remove the threat of poverty. Some 13 per cent of EU households in which at least one member is in employment live below the poverty line (i.e. earn less than 60 per cent of their country's median household income). This figure ranges from 18 per cent of all working households in Portugal and 17 per cent in Spain through to 11 per cent in the UK and Belgium to 7 per cent in Denmark. For a significant share of EU households, in consequence, insertion into the formal labour market does not mean an escape from poverty.

The lesson is that activation policies that insert the unemployed into the formal labour market merely shift the problem from one of the jobless poor to one of the working poor. For example, Oxley (1999) identifies the continuously poor by defining the poverty line to be where annual incomes fall below 50 per cent of the median of household disposable income adjusted for household size. He finds that although just 9 per cent of the continuously poor lived in working households in the UK and 20.9 per cent in Germany, working households comprise 63.6 per cent of the continuously poor in Canada and 48.7 per cent in the United States. The lesson for the EU from North America, therefore, is that activation policies do not resolve poverty. Indeed, and as will be shown in Chapter 6, there is even evidence that insertion into low-wage jobs seems to increase poverty. Measured in terms of the capacities of households to meet needs and desires, it reduces the ability of people to either meet needs themselves or draw upon informal support networks. Whether government measures such as tax credit systems to provide a basic minimum wage for the employed can reduce the poverty witnessed by the working poor, moreover, will depend upon the level at which such credit is set.

Attitudes towards employment

It is not only the existence of the working poor, however, that calls into doubt the desirability of pursuing full-employment as the principal solution to poverty and exclusion. Throughout the advanced economies, there is a deep division between government policy and individual attitudes. In policy-making circles, the desire seems to be to make employment the central focus of people's lives. Indeed, many on the Left who quite rightly used to bemoan the exploitation inherent in employer-employee relationships and profit-motivated exchange relations are now amongst the principal advocates of inserting people precisely into this relationship. How this is meant to represent a path that transcends capitalism and socialism is hard to decipher, at least so far as the capitalism side of the coin is concerned.

Similarly, many 'old style' feminists who were quick to point out the dangers of patriarchal subjugation when women were confined to the home seem quite happy for their sisters to enter subjugation under capitalist social relations by advocating employment as the principal route to their liberation (e.g. McDowell 2001) as they equate, or maybe confuse, this with financial independence from men. Few stop to consider whether advocating the greater subjugation of women by capitalist social relations, especially given the low-paid nature of women's employment, is the route to their liberation. Perhaps this is because these analysts are so immersed in their own 'careers' that they do not consider it may not be the same for others. Perhaps it is because they themselves are the employers of their sisters as cleaners and nannies. Perhaps, however, it is because they can see no feasible alternative future beyond a commodification of economic life. If this is indeed the case, then such academics, like their political counterparts, perhaps need to start to listen to the population at large.

At the very moment that the 'employment ethic' has moved to centre stage in both policy-making circles and academic theories, it has become evident that many people are starting to redefine the importance of employment in their everyday lives (e.g. Cannon 1994, Coupland 1991, Franks 2000, Gorz 1999, Maffesoli 1996, Schor 1991, Zoll 1989). Coupland (1991), for example, highlights the existence of Generation X who refuse to be 'dead at 30, buried at 70'. Like the young Germans studied by Zoll (1989), Generation X will not settle into any of the occupations for which they are suited because none of these has 'sufficient substance', none are sufficiently worthwhile. They prefer to 'hang loose', drifting from one temporary 'McJob' to another, always retaining as much time as possible to follow the favoured activities of their tribe (Maffesoli 1996).

A few years later, this account of Coupland (1991) was complemented by an international survey of young graduates in North America, Britain and the Netherlands. As Cannon (1994: 13) concludes,

> 'Generation X' no longer define themselves through reference to their employment. They have a personal agenda that is more important than that of the organisation they work for, and they may be motivated by a sense of ethical value or genuine social utility – a 'worthwhile ethic' rather than a 'work ethic'. They value their autonomy, rank 'greater control over their time' as the third most important priority behind money and the utilising of their intellectual faculties, and they 'want a better balance between work and other life interests – hobbies, leisure activities, and in the future time spent with the family'.

Two further surveys conducted in France on young graduates of the Grandes Ecoles confirm in every respect these conclusions. Cited by Gorz (1999: 62), a 1990 survey showed that 'what comes out way ahead of every-thing else is the possibility of working when it suits them, so as to be able to devote more time to personal activities'. In 1993, moreover, a survey of current and past students at the prestigious Ecole Polytechnique confirmed this disaffection regarding careers and the general preference for multi-activity and part-time working. As Gorz (1999: 62) summarises, 'the relation to work [employment] is growing looser because life goes on elsewhere' and particularly in 'unpaid activities which are regarded as socially useful'.

There is thus a marked de-centring of the idea about the growing involve-ment and identification of the whole person with his/her job. This disaffection with employment is spreading in all countries and throughout the entire working population, however obsessive the concern with income and the fear of losing one's job. In Germany, only 10 per cent of the working popula-tion regard their employment as the most important thing in their lives. In the US, the proportion is 18 per cent, as against 38 per cent in 1955 (Gallup

Monthly September 1991, cited in Gorz 1999: 63). Amongst Western Europeans aged between 16 and 34, 'work' or 'career' trail far behind five other priorities in the list of 'things which are really important to you personally' (Yankelovich 1995). The five priorities are: having friends (95 per cent), having enough free time (80 per cent), being in good physical shape (77 per cent), spending time with one's family (74 per cent) and having an active social life (74 per cent). Only 9 per cent of those questioned (and 7 per cent of young people between 13 and 25) cited work as 'the main factor for success in life' (Sue 1995). The gulf between 'work' and 'life' thus seems greater than ever: 57 per cent of Britons, for instance, 'refuse to let work interfere with their lives' as against only 37 per cent of those aged between 45 and 54 (cited in Pahl 1995). In a sample of upper middle class full-time employees in the US, meanwhile, Schor (1991) finds 73 per cent take the view they would have a better quality of life if they worked less, spent less and had more time for themselves. The outcome is that some 28 per cent of those questioned had indeed chosen to 'downshift' (i.e. voluntarily earn and spend less) in order to lead a more meaningful life.

It seems, therefore, that a cultural turn has taken place and that much of the political (and academic) world has not caught up with it. People are in employment because the danger is that, if they lose it, they will lose their income and all the opportunities for activity and contact with others. Employment is valued therefore not for the satisfactions that the work itself brings, but for the rights and entitlements attached to employment and employment alone (Gorz 1999). Whilst citizens' rights remain confined to employment rights this situation seems likely to continue.

This is reinforced in evidence on the motives of US women workers. Putnam (2000: 197) asserts that 'virtually all the increase in full-time employment of American women over the last twenty years is attributable to financial pressures, not personal fulfilment'. Examining the DDB Needham Life Style data, women were asked whether they are full-time employees, part-time employees or full-time homemakers. Those employed full- or part-time were then asked whether they work primarily for personal satisfaction or primarily for financial necessity. Those who are full-time homemakers were asked whether they stay at home primarily for personal satisfaction or primarily to take care of children. People have complex feelings about work and the simple 'choose to/have to' dualism fails to capture these fuzzy, mixed and complex motives and attitudes. In the real world, such decisions are doubtless a mixture of all these motivations and others. Nevertheless, as a crude first cut, the standard question distinguishes between women who are working (or not working) mainly because they want to and those who are working (and not working) mainly because they must. Table 1.4 displays the results.

Over the last two decades, 31 per cent of women have taken full-time employment out of financial necessity. This is misleading because their numbers doubled from 21 per cent of all women in 1978 to 36 per cent in

Table 1.4 Employed out of choice or necessity among American women, 1978–99

% of all women	Out of financial necessity (/kids)	For personal satisfaction
Employed full-time	31	11
Employed part-time	11	10
Homemakers	8	29

Source: Putnam (2000: 197).

1999. The proportion of all women working full-time out of personal satisfaction, meanwhile, has not changed over the past two decades, meaning that of all women employed full-time, the share doing so for financial reasons has risen from two-thirds to more than three-quarters.

For part-timers, the same number work part-time out of choice or necessity but over time, there is a modest increase in those working part-time for financial gains rather than personal satisfaction. Some 8 per cent of all women are homemakers who stay at home for childcare reasons. This declined from 11 per cent in 1978 to 7 per cent in 1999. Women who stay at home for reasons of personal satisfaction have fallen from 37 per cent of all women in 1978 to 23 per cent in 1999. These represent women at different stages of the life cycle. Stay-at-home mothers are ten years younger than the national average, while personally satisfied homemakers include a large number of retired women, and this category is ten years older than the national average (Putnam 2000).

In the UK context, the controversial figure of Hakim (2000) similarly explores work-lifestyle preferences amongst men and women. Table 1.5 presents her results. This divides men and women into three broad categories: 'home-centred' people for whom family life remains their main priority throughout life, 'adaptive' people who seek to combine work and family and 'work-centred' people whose main priority in life is their employment or similar activities. For Hakim (2000), there is a strict gender division in terms of the work-lifestyle preferences of men and women. While most women tend to be adaptive, men tend to be more employment-centred. Whatever the gender divisions, the important point that Hakim (2000) reinforces in this table is that employment is not always centre-stage and that many people are seeking a better balance between employment and family life (see also Mauthner *et al.* 2001).

In sum, it is often taken for granted amongst many academics and policymakers that the role of government is to provide people with the opportunity to enter formal employment if they wish to do so. However, it appears that for many people, real choice lies not in the right to employment but in the opportunity to choose between formal and informal work, to combine them in ways that suit them rather than the market. The only conclusion that can be reached therefore is that there needs to be a reconsideration of whether

Table 1.5 A classification of work-lifestyle preferences in the twenty-first century

Home-centred	Adaptive	Work-centred
20% of women 10% of men	60% of women 30% of men	20% of women 60% of men
Children and family life remain the main priorities throughout life	Diverse group including those who want to combine work and family, plus unplanned and unconventional careers, drifters and innovators	Main priority in life is employment or equivalent activities (e.g. politics, sport, art)
Prefer not to engage in competitive activities in the public domain	Want to work, but not totally committed to work career	Committed to employment or equivalent activities in the public domain
Qualifications obtained for intellectual value, cultural capital or as insurance policy	Qualifications obtained with the intention of working	Large investment in qualifications for employment or other activities, including extra education during adult life
Responsive to family and social policy	Very responsive to all policies	Responsive to employment policies

Source: adapted from Hakim (2000: Tables 1.1 and 9.1).

this 'employment ethic' should remain at the heart of economic and social policy. Just as people are recognising and valuing work beyond employment, there appears to be a case for economic and social policy following suit. How this might be achieved will be returned to in Part III of the book.

CONCLUSIONS

In this chapter, the first rationale for adopting an alternative third way approach that seeks to develop the capabilities of people to participate in informal economic activity as a route out of poverty and a means of achieving a more inclusive society has been outlined. Here, the need to pursue such an approach has been explained in terms of the problems surrounding the use of employment as the principal route out of poverty. Our argument has been that pursuing full-employment is neither feasible nor desirable. First, we have identified a significant gap between the current employment situation and a full-employment scenario, despite all of the efforts in advanced economies to achieve the goal of full-employment. The result is that there appears to be little hope, either in the near future or even beyond, that the advanced economies will get anywhere close to a state of full-employment. Given that employment creation is seen as the principal route out of poverty in most advanced economies, the issue that arises is whether such a heavy emphasis should be put on job creation alone to solve the problems of poverty and social exclusion.

Second, and interrelated to this, we have displayed the accumulating evidence that although populations know that employment is currently the principal route out of poverty, many people value the income from employment and the social rights attached to it rather than employment itself. Indeed, when attitudes towards employment are analysed, there appears to be a decentring of employment even if there is a centring of income in many people's lives. This raises the spectre that despite the centring of employment in economic and social policy, attitudes towards employment are shifting in the opposite direction. There thus appears to be a case for economic and social policy following the lead of the population.

2 The informalisation of the advanced economies

For many who believe in a return to the 'golden age' of full-employment, the story of economic development is one in which there is a natural and inevitable shift of work from the non-market to the market sphere as societies become more 'advanced'. In this 'formalisation of work' thesis (see Williams and Windebank 1999a), a linear and uni-dimensional trajectory of economic development is thus envisaged. It is linear because all economies are assumed to formalise. It is uni-dimensional because 'progress' is in only one direction with countries placed at different points on a continuum according to the extent of their formalisation. Minority World, or Western, economies are thus considered 'advanced' due to their supposedly greater formalisation of economic life. Majority World countries, meanwhile, are considered 'under-developed' or 'developing' because of the persistence of 'traditional' informal economic activities, which are seen as a manifestation of 'backwardness' that will disappear with economic 'advancement' and 'modernisation' (e.g. Rostow 1960). The result of such a thesis is that the future becomes no longer open. The future lies in a formalisation of work.

Our contention here, however, is that this story masks its own normative view of economic development and economic history in a discourse of inevitability. Rather than viewing differences in the degree of formalisation between nations, regions and localities as evidence of varying development paths and/or contrasting spatial logics of work, this perspective instead organises these current differences into a temporal schedule. By placing the Majority World behind the West, a single linear history is conceptualised with some nations behind and some in front. This denies nations not only at the back of the queue, but also at the front, having different trajectories to those told in this supposedly universal story. As Gershuny (2000: 21) asserts, however, 'there are no grounds for suggesting that there is any sort of "hidden hand" pointing us to any single, inevitable, and optimal development path'. The aim of this chapter is to start to unpack this dominant discourse in order to call it into question. In so doing, our intention is to evaluate critically not only the 'naturalness' of markets but also their hegemony so as to start to raise the spectre that there may be alternative development paths.

THE FORMALISATION THESIS

Every society must produce, distribute, and allocate the goods that people need to live. As such, all societies have an economy of one sort or another. Economies, however, can be variously organised.[1] In much of the literature, a key distinguishing feature of the present form of economic organisation is that the production, distribution and allocation of goods and services is being commodified or formalised (e.g. Castells 1996, Keck and Sikkunk 1998, Lee 1999, 2000b, Mckay 1998, Polanyi 1944, Scott 2001, Smith 2000, Thompson 1991, Thrift 2000, Watts 1999). By commodification or formalisation is meant the process by which 'goods and services . . . are [increasingly] produced by capitalist firms for a profit under conditions of market exchange' (Scott 2001: 12). As such, economic life in formalised economies possesses three distinct characteristics. First, goods and services are produced for exchange. Second, exchange is monetised and third and finally, the exchange of goods and services on a monetised basis is motivated by the pursuit of profit.

This formalisation thesis that goods and services are being increasingly produced for exchange under market-like conditions in order to generate a profit is a widely held view. Indeed, it lies at the heart, albeit implicitly, of much economic thought. For example, the vast majority of academic enquiry focuses upon the formal sphere, yet explicit rationales to justify this focus are notable by their absence. One can thus only assume that it is because most academic enquiry accepts the formalisation thesis, preferring to focus upon what it views as the growing or dominant formal sphere rather than what it sees as a residual or diminishing informal sphere. Indeed, the hegemony of this view is perhaps most clearly portrayed when examining how those studying the informal sphere often justify their enquiry. Studies of informal childcare, care for the elderly and the gender division of domestic labour, for example, frequently justify themselves in terms of how these informal activities represent a barrier to participation in formal economic activity. In other words, while those studying the formal sphere find it unnecessary to defend their focus, those researching informal work are drawn to justifying themselves in terms of its importance for formalisation. Formalisation itself, therefore, is seldom questioned. Indeed, who would do so? To question the hegemony of what are essentially capitalist social relations seems a fruitless task. The widespread consensus, after all, is that capitalism is victorious, penetrating, expansive and all-powerful (see Gibson-Graham 1996).

Indeed, even when the formalisation thesis is explicitly addressed, the only debate is whether this commodification/formalisation process is near enough complete (e.g. Thrift 2000) or whether it is a slow and uneven process (e.g. Gough 2000, Lee 1999, Watts 1999). Few question formalisation itself. For some analysts, that is, formalisation has stretched its tentacles so far that few areas remain that have not been incorporated under its wing. As Thrift (2000: 96) puts it, 'What is certain is that the process of commodification has reached into every nook and cranny of modern life.' Others, however, although not questioning the overall formalisation process, believe that formalisation is as yet far from total. For them, formalisation is a slower and more uneven process (e.g. Lee 1999, 2000b, Thompson 1991, Watts 1999), albeit still one that is colonising all spheres of life. As Gough (2000: 17) puts it,

> Following decolonisation and the collapse of state socialism at the end of the 1980s, few areas of the world remain to resist the logic of capitalist markets and economic enterprises. This in turn is imposing the needs of capital in more and more areas of life and is weakening the resources of states and citizens to fight back.

Whether the 'colonization of the lifeworld' (Habermas 1975) by formalisation is total or not, however, is not the important issue here. The salient point is that wherever the spotlight shines, the near universal picture being painted is one of formalisation.

Yet one of the most disturbing issues is that hardly any evidence is ever provided to show how more of economic life is mediated through the market. Despite the deluge of references to how 'markets are subsuming greater portions of everyday life' (Gudeman 2001: 144) and how, although 'commodification is not complete . . . the reality of capitalism is that ever more of social life is mediated through and by the market' (Watts 1999: 312), data to support this stance is notable by its absence. Carruthers and Babb (2000: 4), for instance, despite making the sweeping assertion that there has been 'the near-complete penetration of market relations into our modern economic lives' provide no evidence beyond asserting that 'markets enter our lives today in many ways "too numerous to be mentioned"'. Even when evidence is provided, it is often little more than bland statements about the rarity of subsistence economies (Watts 1999) or how commodified *ways of viewing* are pervading more areas of life. Perhaps the evidence is so obvious to all that there is little need to provide it. Perhaps, however, it is not. For example, it is clearly not the case that the advent of commodified interpretations of social life (e.g. the 'marriage market') signifies that those realms of social life *are* now commodified.

This lack of evidence on formalisation and its uneven contours is unacceptable. No other idea in the social sciences is endorsed without detailed scrutiny and there is no reason why formalisation should be an exception. Our aim here, therefore, is to evaluate critically the advance of formalisation. To do this, we seek to answer a number of questions. How deeply has formalisation penetrated life in the advanced economies? Is the trend towards ever more formalised economies? Do informal spheres remain? If so, what forms do they take? And what do we know about where they are to be found and who engages in them? To answer these questions, we here review the secondary data on the extent to which informal modes of production exist in the advanced economies.

Table 2.1 Forms of economic activity

	Monetised	
Exchange	(a) Non-profit motivated	(c) *Non-monetised*
	(b) Profit motivated	
Non-exchanged work	(d) Self-provisioning	

A CRITICAL EVALUATION OF THE FORMALISATION THESIS

As stated above, those who discuss formalisation have made little attempt to show the extent or nature of its penetration. Here, therefore, the idea that goods and services are increasingly produced for monetary exchange under profit-motivated market-orientated conditions (i.e. the formalisation thesis) is put under the microscope. If it is correct that formality is increasingly colonising and penetrating social life then the shaded area of Table 2.1 should be expanding as it colonises all of the non-shaded areas. To test whether this is the case in the context of the advanced economies, first, we analyse the amount of unpaid work (i.e. non-exchanged work and non-monetised exchange) relative to paid work and second, we find whether there is monetised exchange outside the logic of profit. This will provide a measure of the extent to which formalisation has colonised economic life and will pinpoint whether there are alternative economic spaces of self-provisioning, exchange grounded in non-monetary relations and monetary exchange where the profit motive is absent.

The persistence of non-exchanged work and non-monetised exchange

It might be assumed that most economic activity in advanced economies occurs in employment-places such as the factory, office and retail outlet. Indeed, given that most economic data gathered in the advanced economies measures only the magnitude and character of formal economic activity, reflecting the dominance of the ideology that the formal sphere is the principal mode of economic organisation in contemporary society, such a view is seldom questioned.

However, lurking in the shadows is a long tradition that has sought to incorporate unpaid work into economic statistics. One of the pioneers of this approach was Margaret Reid, who in her 1934 book *Economics of Household Production* (Reid 1934), expressed concern about the exclusion of domestic production from national income accounts and designed a method to estimate the value of home-based work. It was only in the latter half of the twentieth century, however, that such a view started to take-off. From the late 1960s, a vast social science literature, heavily influenced by the women's movement, began to call for the recognition and revaluing of unpaid work (see Beneria 1999, Delphy 1984, Gregory and Windebank 2000, McDowell 1991). One aspect of this was a campaign to change the ways in which international bodies and national governments measure economic activity (see Beneria 1999, Chadeau and Fouquet 1981, James 1994, Luxton 1997). Arguing that the national accounts are systematically skewed because they ignore the value of women's unpaid work, a campaign ensued to get unpaid work adequately recognised in national accounts.

This has met with considerable success. Of potential long-term significance is a recommendation, made in 1993 by the United Nations (UN) after a review of the UN System of National Accounts, that 'satellite national accounts' be developed that incorporate the value of unpaid work. This became an obligation for many advanced economies under the terms of the Final Act of the 1995 UN Fourth World Conference on Women in Beijing (UN 1995, Section 209: f, g). These computations are an adjunct to the standard national accounts that were roundly criticised for their silence on women's unpaid work. Although such accounts remain subsidiary, it is clear that they represent a step in the right direction. Indeed, both the European Commission and many national governments have developed, or are developing, such 'satellite' accounts that measure the value of unpaid work.

In theory, the size of unpaid relative to paid work could be measured by examining the volume and/or value either of the inputs or outputs of each type of work (Goldschmidt-Clermond 1982, Gregory and Windebank 2000). In practice, the way governmental agencies have taken on board such recommendations is by measuring the volume and/or value of the inputs. To do this, time-budget studies have been used (e.g. Murgatroyd and Neuberger 1997, Roy 1991). A time-budget study is a technique for data collection whereby the research participants complete diaries chronicling the number of minutes spent on a range of activities. From these, it is possible to calculate the time spent on different forms of work. Indeed, it is now widely accepted that measuring time use is as useful and accurate in assessing informal work as money is in measuring paid employment (Gershuny and Jones 1987, Gershuny *et al.* 1994, Juster and Stafford 1991).

Table 2.2 displays the results of these time-budget studies. This reveals some profound findings about economic life and the trajectory of economic development. First, these studies reveal that when the amount of time that people spend engaged in unpaid work is calculated, this occupies 44.7 per cent of all working time. The tentacles of formalisation, therefore, do not appear to have extended as far as many previously imagined. The sphere of unpaid work, so long considered the residual and diminishing 'other', is of a similar magnitude to the paid sphere, measured in terms of the volume of the time spent on it. Economic life, in consequence, appears to be far from totally formalised. Indeed, in many nations (e.g. Canada,

Table 2.2 The trajectory of economic development: unpaid work as a percentage of total work time, 1960–present

Country	1960–73	1974–84	1985–present	Trend
Canada	56.9	55.4	54.2	Formalisation
Denmark	41.4	—	43.3	Informalisation
France[a]	52.0	55.5	57.5	Informalisation
Netherlands	—	55.9	57.9	Informalisation
Norway	57.1	55.4	—	Formalisation
UK	52.1	49.7	53.9	Informalisation
US[b]	56.9	57.6	58.4	Informalisation
Hungary	53.6	46.8	—	Formalisation
Finland	—	51.8	54.5	Informalisation
20 Countries	43.4	42.7	44.7	Informalisation

Sources: a) Chadeau and Fouquet (1981), Roy (1991), Dumontier and Pan Ke Shon (1999); b) Robinson and Godbey (1997). Other countries derived from Gershuny (2000: Tables 7.6, 7.12, 7.16).

France, Netherlands, UK, US), the time spent on unpaid work is greater than the time spent in employment.

The second profound finding portrayed in Table 2.2 concerns the trajectory of economic development. It reveals that the shift of work from the unpaid to the paid sphere has not only stalled over the past forty years but has even gone into reverse in some countries. In nations such as Denmark, Finland, France, the UK and the US, unpaid work has occupied an increasing proportion of people's total working time. This, however, is not due to an absolute growth in the time spent on unpaid work. The total number of hours spent on such work has declined. But the time spent in paid work has decreased faster than the time spent in unpaid work. The result is that over the past four decades, there has been an informalisation of economic life.

This informalisation of economic life, however, is not perhaps as surprising as it first appears. The origins of the formalisation thesis lie in Karl Polanyi's seminal book, *The Great Transformation*, which identified the shift from the non-market to the market sphere (Polanyi 1944). In this book, he went to great lengths to point out that this was merely a shift in the balance of economic activity from the non-market to the market sphere. Even if some have since portrayed this transformation as rather more complete than Polanyi ever wished to suggest (e.g. Harvey 1982, Thrift 2000), the time-budget data in Table 2.2 display that Polanyi was quite correct not to over-exaggerate the reach of the market. Indeed, what appears to have been happening over the past forty years is that a so-far unidentified but profound second 'great transformation' has taken place in many advanced economies. Measuring the time spent working in the unpaid and paid spheres, there has been a shift from the market to the non-market sphere.

Indeed, these time-budget studies may well be underestimating the degree to which work has been informalised. First, this is because the time spent in paid work is overestimated in time-budget studies. People tend to count the total time spent in activity associated with the employment-place as time spent in paid work when much of this time may include meal- and coffee-breaks, associated travel as well as socialising. Second, this is because the time spent in unpaid work is underestimated in time-budget studies. First, many time-budgets only measure an individual's commitment in time to a concrete activity. They do not assess the time and effort involved in planning and managing one's own and others' activities. This often occurs when one is watching television, lying in bed or undertaking some other supposed leisure pursuit or indeed when one is engaged in one's employment (Haicault 1984). In addition, although time-budget studies measure the time spent in unpaid work, such as caring for one's family, they do not assess the emotional and affective activity involved in family work which either is ignored completely, or indeed, is portrayed as leisure and socialising. Nor do they capture the way in which much of the time not actually spent in the service of the family is still constrained time (Chabaud *et al.* 1985). Third and finally, these time-budget studies have not so far differentiated monetised exchange into its formal and informal components. This, however, is important. Paid informal work grew from 5 per cent of the GDP of the EU in the 1970s to between 7 and 16 per cent of GDP during the 1990s (European Commission 1998a). If this was incorporated into the equation, and as will be returned to below, then the formal sphere would be smaller than is suggested in Table 2.2.

Reflecting the fact that the formal sphere is the benchmark against which the importance of unpaid work is measured, attempts are frequently made to put a monetary value on this activity. To do this, three techniques have been adopted: opportunity costs, housekeeper wage costs, and occupational wage costs. In each, monetary values from the market sector are used to impute values to such activity or its products (Luxton 1997). First, the opportunity-costs model calculates the income the worker would have earned if s/he had been in the paid labour force instead of undertaking unpaid activity. Second, the housekeeper wage-costs approach calculates how much a worker in paid employment doing similar work is paid. Third and finally, the occupational wage-costs approach measures the price of household inputs by calculating market equivalents for the costs of raw materials, production and labour and comparing it with market prices for each product and/or service. An extensive literature assessing the various approaches exists and methods have become increasingly sophisticated (Bittman 1995, Ironmonger 1996, Jenkins and O'Leary 1996, 1997, Lutzel 1989, Luxton 1997, OECD 1997).

The results of putting a monetary value on this work make interesting reading. In the UK, the ONS survey conducted in 1995 finds that using replacement costs whereby time spent cooking is replaced by cooks and childcare with childminders, unpaid work was worth 56 per cent of GDP because these sectors tend to be the lowest paid in the economy. However, valuing the time spent on unpaid work at the same average wage rate as paid employment as a whole it would be 122 per cent of GDP (Murgatroyd and Neuburger 1997).

However, one issue with all of these valuation methods is that they accept existing gender divisions of labour and pay inequalities as normal and unproblematic. The result is that the valuations produced serve to reinforce gender and broader social inequalities rather than challenge them (Bryson 1996, Chadeau and Fouquet 1981). For example, childcare is usually assessed at (low-paid) childcare worker rates rather than, for instance, at rates earned by psychologists, teachers and nurses with no questioning of why such childcare workers are so lowly paid. The result is that even if more time were spent on informal than formal work, when evaluated in monetary terms, it would be worth less.

Moreover, there are inherent dangers in putting a price on such work. Enumerating a financial price on this effort simply converts everything into a purely market transaction and makes it part of the formalised accountancy culture. But it seems inevitable that the only way that domestic effort can be assigned any social status is through financial value because that is what a market-driven society dictates. It is not only unpaid domestic work that raises these paradoxes. The whole sector of voluntary work and caring confronts the same problem. Nevertheless, at least these measures provide some indication of the amount of non-exchanged work in particular and unpaid work more generally that takes place in the advanced economies.

Time-budget studies, however, are not the only measure of the amount of non-exchanged work and non-monetised exchange in these economies. A voluminous literature on the gendered aspects of unpaid work displays the vast amount of non-exchanged work taking place and how most of this work is undertaken by women (e.g. Gardiner 1997, Gregory and Windebank 2000, Himmelweit 2000). There is also an extensive literature on various aspects of non-monetised exchange such as gift giving, voluntarism and unpaid reciprocity (e.g. Berking 1999, Caplow 1982, Cheal 1988, Corrigan 1989, Putnam 2000, Wuthnow 1997). Wuthnow (1997), for example, points out that in America, 80 million Americans (45 per cent of the population aged over 18) spend five or more hours each week on voluntary services and charitable activity, which adds up to well over $150 billion a year. In the EU in 1999, meanwhile, the European Commission (2001b) shows that three out of ten people spend time helping people on a voluntary basis. There is thus a solid empirical base that challenges the view that 'Monetary relations have penetrated every nook and cranny of the world and into almost every aspect of social, even private life' (Harvey 1982: 373). A further interesting question, however, is whether all paid work is economic activity conducted under profit-motivated market-based exchange relations.

Monetised exchange beyond the profit motive

The view that monetary exchange is always profit-motivated runs deep in both neo-classical and Marxian economistic discourse. The common assertion is that as monetised exchange has penetrated more areas of social life, this has marched hand in hand with the profit motive. As such, the only kind of monetary exchange that is seen to exist is profit-motivated exchange. As Sayer (1997: 23) argues,

> The commodity may be valued by the user for its intrinsic use value, but to the seller it is unequivocally a means to an end, to the achievement of the external goal of making a profit, and if it is unlikely to make a profit it will not be offered for sale.

Here, however, the question is raised of whether monetised exchange is always profit motivated. In recent years, this has been challenged (e.g. Crang 1996, Crewe and Gregson 1998, Davies 1992, Gertler 1997, Gudeman 2001, Lee 1996, 1997, 2000a, 2000b, Thrift and Olds 1996, Zelizer 1994).

On the one hand, the cultural turns across the social sciences have resulted in a widespread questioning of this narrowly economistic way of viewing exchange (see Crang 1996, Crewe and Gregson 1998, Davies 1992, Lee 2000a, Thrift and Olds 1996, Zelizer 1994). There is a strong anthropological tradition that sees exchange mechanisms in advanced economies as less 'embedded' than those in pre-industrial societies. Mauss (1966), for instance, perceived Western developed exchanges as 'thinner', less loaded with social meaning and less symbolic than traditional exchanges. In this view, traditional exchange is a socially embedded process that obeys a different logic to the economic rationality and profit maximisation of exchange in more advanced economies. As Crewe and Gregson (1998: 41) incisively point out, however, 'the major defect of such market-based models of exchange is simply that they do not convey the richness and messiness of the exchange experience' in the advanced economies.

In recent years, in consequence, this profit-motivated view of monetised exchange has been contested. Drawing upon the substantivist anthropology of Polanyi (1944), there has been considerable criticism of the 'formalist' anthropology approach that assumed price-fixing and profit-motivated markets to be the universal economic mechanism. For Polanyi, the error was in equating the human economy in general with its market form. In positing the 'great transformation', he suggested that there had been a move from socially embedded economic exchange, often based on reciprocity, to the impersonal and placeless operation of the market. However, Polanyi also correctly asserted that the market, whatever forms it takes, is itself a social product. In the social sciences, this has been widely recognised in the writings of those influenced by the 'cultural turn/s'. These analysts have sought to unpack the nature of monetary exchange so as to rework the social nature of the economic (e.g. Crang 1996, Crewe and Gregson 1998, Lee 1996, 1997, 2000a, 2000b) and, in so doing, show that alternative social relations, motives and pricing mechanisms exist.

It is not only how one looks at monetary exchanges, however, that influences whether one finds the ubiquitous presence of the profit motive. It also depends on where one looks. Up until now, the lens of the social sciences has been firmly focused upon formal economic spaces. Few have cast their net more widely. When they have done so, however, there have been some key insights into the relationship between monetary exchange and the profit motive. First, studies of alternative economic spaces such as the garage sale (Soiffer and Herrman 1987), the car boot sale (e.g. Crewe and Gregson 1998) and charity and second-hand shops (e.g. Gregson and Crewe 2002) have highlighted forms of sourcing, commodity circulation, transaction codes, pricing mechanisms and value quite different from those that typify profit-motivated market-orientated exchange. Second, a burgeoning literature on local currencies reveals how it is possible for monetary exchange to take place under alternative social relations and for motives other than profit (e.g. Lee 1996, North 1999, Offe and Heinze 1992, Williams *et al.* 2001).

Third, studies of paid informal exchange in UK urban areas reveal how customers and suppliers conduct this work largely for non-profit-motivated reasons (e.g. Williams and Windebank 2001a, 2001c). Fourth and finally, investigations of how people spend their money have displayed the widespread existence of non-profit-motivated rationales. These range from studies showing how the celebration of Christmas accounts for about 4 per cent of total annual expenditures (Berking 1999, Caplow 1982) to Eurobarometer surveys of how three out of five people gave money or goods to help people living poor or socially excluded lives during 1999 (European Commission 2001b). Consequently, there is a growing literature on how monetised exchange is not always profit-motivated. Indeed, recent studies even suggest that monetary exchanges in some formal economic spheres are not always motivated purely by profit (e.g. Lee 2000a, Zafirovski 1999).

Explaining the persistence of informal work

Here, therefore, it has been shown that informal work persists in the advanced economies. There exist goods production and service provision that is not exchanged, non-monetised exchange persists and monetary transactions not imbued with the profit motive prevail. How, therefore, can the perseverance of these informal modes of production in the advanced economies be explained? Sometimes, they have been seen as vestiges of a pre-capitalist past awaiting incorporation. More usually, they have been explained as the product of a new post-Fordist regime of accumulation that is offloading social reproduction functions from the formal sphere back on to the informal sphere (e.g. Castells and Portes 1989, Lee 1999, Portes 1994). In this view, the breakdown of the post-war economic regulations and welfare state through a general trend of deregulation and flexibilisation of social relations of production, and the transferring of social services to private and communal hands (Gershuny and Miles 1983, Pahl 1984) has led to the informalisation of economic life. Informal modes of production have thus expanded to occupy spaces of production (and reproduction) previously covered by market relations and state subsidies.

In this perspective, it is the contradictions inherent in the formalisation process that have led to the informalisation of some spheres of social reproduction. In order to compete in the global commodified economy, for example, many advanced economies have had to reduce social costs (see European Commission 2000a, 2001b). To achieve this, activities associated with social reproduction have been taken out of the formal sphere and imposed once again on the informal sphere, resulting in a shifting balance between the formal and informal modes of production.

Although such structural 'economic' explanations are necessary and perhaps even sufficient for understanding the persistence and even resurgence of informality, the fact that informalisation can also be read in 'cultural' terms cannot and should not be ignored. Rather than view informality as purely imposed on people by structural economic constraints, it can be read to be a result of agency. It is increasingly apparent that, confronted by dissatisfaction in their formal employment, many people view informal activity as a source of work satisfaction, pleasure and a means of individualising the products of consumer society for their own purposes (e.g. Gorz 1999, Williams and Windebank 2001b). As formal employment becomes more intense, therefore, individuals are not externalising their domestic activities and reducing their reciprocal obligations but, instead, are seeking solace in this realm in order to get the pleasure that they cannot find in their formal work. Indeed, this perhaps explains why multiple-earner households throughout the advanced economies conduct more unpaid domestic work, community work and paid informal work than their jobless counterparts (see Williams and Windebank 1993, 2000a, 2000b). Indeed, viewing such informal economic spaces as sometimes 'chosen spaces' reinforces the view of Urry (2000: 146) that 'a largely unintended effect of a highly individualised and marketised society has been the intensification of social practices which systematically "evade the edicts of exchange value and the logic of the market"'.

It is important, in consequence, not to see informal work spaces purely as a by-product of structural economic transformations. They also need to be read more positively as sites of resistance to the logic of formalisation. Viewed in this way, these sometimes chosen spaces not only need to be given symbolic value as 'spaces of hope' (Harvey 2000) but also need to be read as development sites for the demonstrable construction and practice of alternative social relations and logics of work outside profit-motivated market-orientated exchange.

As will be shown throughout this book, however, informal economic activities cannot be universally explained in the same manner across time and space. Amongst some affluent populations, for example, they can be at times interpreted as 'chosen spaces'. However, this is not largely the case amongst deprived populations where they are much more accurately explained in structural terms. To start to see this, we here begin by reviewing the literature on the socio-spatial variations in the magnitude and character of informal work.

SOCIO-SPATIAL VARIATIONS IN INFORMAL WORK

In Part II of this book, the detailed findings will be reported on who conducts informal work in contemporary England and whether lower-income areas conduct more informal work than higher-income localities. Here, however, we explore the findings of studies that have been conducted in various other nations, regions and localities throughout the advanced economies in order to map out current understandings of its socio-spatial variations. To bring some order to the diverse findings that have been produced when researching this issue, it is first necessary to unpack the various tenets of conventional discourse concerning where it takes place, who partakes in such work, its character and why they do it. This will then enable the findings that have challenged these beliefs to be understood.

Conventional discourse on socio-spatial variations in informal work: the 'marginality thesis'

The traditional conceptualisation of informal work, and one that is only slowly receding, is that it is undertaken by those marginalised from employment as a survival strategy (e.g. Button 1984, Gutmann 1978, Matthews 1983, Rosanvallon 1980). This has become known as the 'marginality thesis' due to its views on where it takes place, who does it and why people engage in it.

First, that is, deprived neighbourhoods are viewed as having stronger traditions of informality than more affluent areas (e.g. Button 1984, Elkin and McLaren 1991, Robson 1988). Such work is believed to be spatially concentrated in such a manner because second, marginalised groups are seen to undertake the vast majority of informal work (e.g. Lagos 1995, Maldonado 1995, Rosanvallon 1980). Put another way, the assumption is that informal work is primarily a coping practice of the poor and unemployed. Elkin and McLaren (1991: 217), for example, talk of 'disadvantaged localities where informal work is often very important', whilst Haughton *et al.* (1993: 33) assert that

> the distribution of informal work – may well be especially important in areas of high unemployment, in part acting as a palliative, in part merely recycling people between employment and unemployment and possibly reducing official unemployment statistics.

Inextricably related to the view that informal work is conducted by marginalised areas and groups is the conceptualisation of the motives of participants. Third, that is, informal work is primarily viewed as an economic survival strategy conducted in order to make or save money (Button 1984, Castells and Portes 1989, Gutmann 1978, Isachsen *et al.* 1982, Kesteloot and Meert 1999, Lagos 1995, Maldonado 1995, Matthews 1983, Petersen 1982, Portes 1994, Rosanvallon 1980, Sassen 1989, Simon and Witte 1982).

These conceptualisations of both its nature and the motives of participants gained currency throughout the 1970s and 1980s (Button 1984, Castells and Portes 1989, Gutmann 1978, Isachsen *et al.* 1982, Matthews 1983, Petersen 1982, Rosanvallon 1980, Sassen 1989, Simon and Witte 1982). Indeed, they have remained in circulation to varying extents throughout the 1990s up until the present day (e.g. Kesteloot and Meert 1999, Lagos 1995, Maldonado 1995, Portes 1994). Running parallel to this marginality thesis, however, has been a steady stream of writing that has sought to challenge one or more of its tenets.

Alternative discourses on socio-spatial variations in informal work

In a bid to counter the view of informal work as conducted by marginalised areas and groups for unadulterated economic reasons, there has been a questioning of where such work takes place, who does it and why people engage in it.

Spatial variations in informal economic activity

First, there has been a questioning of where it takes place. With many empirical studies displaying that informal work is not confined to lower-income areas, a more refined understanding has emerged. This argues that despite a 'cocktail' of economic, institutional, social and environmental influences combining variously in different areas to produce particular local outcomes, some strong common spatial patterns can still be identified in most advanced economies (Williams and Windebank 1995b, 1998a, 1998b). In the Netherlands, a study of six localities by Van Geuns *et al.* (1987) asserts that the higher the rate of unemployment in an area, the lower is the level of informal work. In France, meanwhile, studies in both Orly-Choisy (Barthe 1985) and Lille (Foudi *et al.* 1982) indicate that there are poverty black spots in which the unemployed cannot escape from their deprivation through informal economic activity. Studies of more affluent new towns and commuter areas in France (Cornuel and Duriez 1985, Tievant 1982), in contrast, discover a relatively high amount of informal economic activity. In Italy, similarly, studies show that such activity is more extensive in the relatively affluent northern regions than in the more deprived southern regions (Mattera 1980, Mingione 1991). As Mingione and Morlicchio (1993: 424) declare, 'the opportunities of [*sic*] informal work are more numerous, the greater the level of development of the surrounding social and economic context'. Although this finding is by no means universal (see Williams and Windebank 1998a), it is now widely accepted that informal work is not always and necessarily concentrated in marginalised areas as purported in the marginality thesis. However, it is not only where informal work takes place that has been reconfigured.

Socio-economic variations in informal economic activity

Second, and related to this reconceptualisation of the spatial distribution of informal work, there has been mounting criticism of the view that such work is concentrated amongst marginalised groups. This has occurred in nearly all of the major regions of the advanced economies. Take, for example, northern Europe. In the Netherlands, Van Geuns *et al.* (1987) find in all six localities studied that the unemployed generally do not participate in such work to the same extent as the employed. This is reinforced by Van Eck and Kazemier (1985), Koopmans (1989) and Komter (1996). It is also echoed in studies carried out in France (Barthe 1988, Cornuel and Duriez 1985, Foudi *et al.* 1982, Tievant 1982), Germany (Glatzer and Berger 1988) and in Britain (Economist Intelligence Unit 1982, Howe 1990, Morris 1994, Pahl 1984, Warde 1990, Williams and Windebank 1999b, 2001b). All show that informal work is greater amongst the employed rather than the unemployed.

Similar findings are identified in southern Europe in Spain (e.g. Benton 1990, Lobo 1990a), Portugal (Lobo 1990b) and Italy (Cappechi 1989, Mingione 1991, Mingione and Morlicchio 1993, Warren 1994). This reconfiguration is not only prevalent across Europe but also in North America (e.g. Fortin *et al.* 1996, Jensen *et al.* 1995, Lemieux *et al.* 1994, Lozano 1989). The result, therefore, is that the extent of participation across both areas and social groups has been reconfigured to challenge the tenets in conventional discourse about its distribution.

The motives for informal work

Coupled to this rethinking of where it takes place and who does it has been a criticism of the view that it is purely motivated by economic reasons. Recognising that a wider cross-section of society is involved than merely marginalised groups, other intentions have been argued to be involved (Cornuel and Duriez 1985, Jensen *et al.* 1995, Komter 1996, Leonard 1994). On the whole, however, non-economic motives, such as the desire to develop reciprocity and trust, have been mostly assigned to those who are employed, especially those with high formal incomes (see Cornuel and Duriez 1985, Williams and Windebank 1998a). For unemployed and low-income groups, the view persists that their principal motive for participation is to save money

or to earn sufficient income in order to get by (Howe 1990, Jordan *et al.*
1992, Leonard 1994, MacDonald 1994, Rowlingson *et al.* 1997).

CONCLUSIONS

In sum, this chapter has evaluated critically the formalisation thesis, which
asserts that non-capitalist production is disappearing, albeit slowly and
unevenly, and is being replaced by goods and services produced by capitalist
firms for a profit under conditions of market exchange. Examining the
degree to which there exists non-profit-orientated monetised exchange, non-
monetised exchange and non-exchanged work in contemporary society, we
have here shown that even in the heartland of formalisation – the advanced
economies – there are large alternative economic spaces where the logic of
formalisation is absent.

The 'great transformation' from a non-market to a market society, upon
which many of the current assumptions about economic development are
founded, must be thus treated with caution. Examining the evidence of time-
budget studies that measure the amount of time spent engaged in various
forms of work, formal employment does not appear to be the universal means
of getting work done in the advanced economies. Rather, informal modes of
production, so long considered the residual and diminishing 'other', represent a significant means by which work is undertaken and these appear to
be growing in comparison with employment. The profound finding of this
chapter, therefore, is that informal modes of production persist in the
advanced economies. Indeed, nearly half of people's total work time is spent
engaged in informal work and, at least in some advanced economies, there
is a tentative shift in the balance of work towards this sphere. Unlike those
asserting that the logic of the market has penetrated even our most intimate
social relations of family and community (e.g. Thrift, 2000, Wolfe, 1989),
the finding of this chapter, to use the words of Zelizer (1994: 215), is thus
that 'The vision of society fully transformed into a commodity market is no
more than a mirage.'

Nor, moreover, is this persistence of informal work simply due to the social
reproduction of marginal populations being offloaded from the formal to the
informal sphere. Although the conventional view is that this work is concentrated amongst marginalised populations, this chapter has reviewed
the empirical evidence on its socio-spatial distribution. This reveals that, if
anything, such work is more likely to be heavily used by relatively affluent
population groups. Indeed, informal modes of production usually reinforce
rather than reduce the socio-spatial inequalities produced by formal employment. To explain their persistence, we have thus argued that such activities
cannot wholly be seen to be a manifestation of the contradictions inherent
in post-Fordism that result in the social reproduction of marginalised populations being offloaded from the formal sphere back on to the informal sphere.

The greater prevalence of informal work amongst more affluent populations suggests that they might also be sometimes 'chosen spaces'. Put another way, they display the existence of 'cultures of resistance' to the edicts of formalisation that provide a lens through which the demonstrable construction and practice of alternative social relations and logics of work beyond market-orientated exchange and the profit motive can be seen.

Whatever the reason for their persistence, the important finding of this chapter is that informal economic activities are a large and growing segment of economic life, at least measured in terms of the time spent engaged in such work. They thus warrant much greater attention than has so far been the case. Until now, the focus of policy has been upon formal employment in the mistaken belief that this is the largest realm of economic activity and is the form of work that is growing. The other half of the economy has not been subject to policy intervention. However, given both that the socio-spatial inequalities in the formal sphere are mirrored in the informal realm, and that this realm is growing relative to formal employment, there is a pressing need to consider what is to be done about informal work. If coping capabilities are to be enhanced, it is not just the ability of people to participate in the formal labour market that needs to be considered but also their ability to participate in informal work. In the light of this, the next chapter reviews the various possible policy responses that can be adopted towards informal work along with their implications.

Note

1   Most analyses tend to conceptualise three different variants of economic organisation: market, state and community (Boswell 1990, Giddens 1998, Gough 2000, Polanyi 1944, Powell 1999, Putterman 1990, Thompson 1991), although different nametags are often used. For instance, Polanyi (1944) refers to 'market exchange', 'redistribution' and 'reciprocity' whilst Giddens (1998) uses 'private', 'public' and 'civil society'.

3   Discourses on informal work
and their implications

The last chapter revealed that the trajectory of economic development in the advanced economies is not necessarily and inevitably towards formalisation. For some four decades, the informal sphere has been growing relative to employment, at least measured in terms of the time spent on such activity, and as much time is now spent engaged in informal work as in formal employment.

What, therefore, is to be done about this large and growing sphere of informal economic activity? Should attempts be made to repress such work? Should we adopt a laissez-faire approach towards it? Or should we attempt to work with the grain and cultivate such activity? To answer these questions, this chapter deals with each in turn. First, we evaluate critically the implications of seeking to *repress* such work, as advocated by old-style social democrats. Second, we evaluate the implications of adopting a laissez-faire approach, as advocated by many neo-liberals and third and finally, we explore the implications of attempting to *harness* work beyond employment.

The argument of this chapter, as in the rest of this book, is that to alleviate poverty and build a more inclusive society, the capabilities of people to participate in informal work will need to be developed. The first option of repressing informal work is thus argued to be both impractical and undesirable. It is impractical because informal work is deeply embedded in everyday life and the evidence points towards an informalisation rather than formalisation of economic life. It is undesirable because this work is often the preferred means by which people conduct activities and a key ingredient of the social glue that binds communities together. A laissez-faire approach, meanwhile, results in numerous negative consequences. Given that those excluded from the formal labour market are also less able to participate in informal work, a laissez-faire approach towards informal work will merely further consolidate the socio-spatial inequalities produced by employment.

This chapter thus argues that going with the grain and cultivating informal work is the only viable option. The second half of this chapter is then devoted to investigating the consequences of the contrasting approaches towards developing informal work. Distinguishing between the types of work that various analysts wish to cultivate and their reasons for doing so, four contrasting approaches towards cultivating informal work are identified:

- Harnessing third sector associations as a springboard into employment;
- Cultivating the informal sphere as a complementary means of welfare provision;
- Developing informal work as a complementary means of livelihood; and
- Harnessing informal work as an alternative to employment.

The implications of pursuing each of these approaches are considered in turn. The aim in so doing is to begin the process of rethinking the role that informal work could play in paving an alternative third way. This will then set the scene for the rest of the book.

REPRESSIVE DISCOURSES ON INFORMAL WORK

On the whole, repressive discourses (RED) towards informal work are adopted by old-style social democrats. Free market capitalism is seen to produce many problematic effects that can be reduced or tackled by state intervention in the market sphere. The state thus has the obligation to provide public goods that markets cannot deliver or can do so only in a fractured way. A strong government presence in the economy, and other aspects of society as well, is seen as normal and desirable. The result is that social democratic thought stands or falls by its capacity to deliver a society that generates greater wealth than unbridled capitalism and spreads that wealth in a more equitable fashion.

The traditional view of this approach is thus that the state needs to intervene both in formal work and welfare to achieve full-employment and/or a comprehensive and universal formal welfare 'safety net'. As such, the informal sphere has no positive role to play in the future of work and welfare or in tackling poverty and social exclusion. This approach, however, suffers from several inherent problems. These relate first to the possibility of creating a 'full-employment/comprehensive welfare state' scenario and second, to the possibility of eradicating informal work.

A critical evaluation of the repressive approach

This approach believes the future of work to lie in a return to the 'golden age' of full-employment. As Chapter 1 argued, however, the gap is considerable between current employment participation rates and a full-employment scenario and there is little evidence that over time, it is being significantly closed. Indeed, it seems unrealistic to expect any advanced economy to achieve a state of full-employment as understood four decades ago. To achieve this, either a return to outdated gender divisions of labour (with men in employment and women at home) is necessary or the provision of jobs for all men and women desirous of employment is required, which has never so far been achieved. As the European Commission (1996a: 28) conclude, 'it is hardly likely that we will return to the full employment of the 1960s'. It seems therefore that the old-style social democrats dream of a stable full-employment society guaranteeing incomes and social participation to the vast majority of the population is receding ever further from their grasp.

If the advanced economies are not returning to the 'golden age' of full-employment, then the question that begs an answer is whether it is any longer possible to construct a formal welfare 'safety net' to protect those households and workers excluded from employment. To evaluate this, we here consider the social democratic EU rather than neo-liberal welfare regimes such as the US and UK. Even here, however, any review of the direction of formal welfare provision gives little cause for optimism.

The aim of the 1986 Single European Act, and the Single European Market (SEM) in particular, was to revitalise tired European economies, make industry more productive and promote faster European growth. By opening up vastly differing economies more fully to one another, the concern was that there might be a levelling down of social protection. A social dimension to the Single European Act was thus introduced in the form of the Social Charter. At the outset of discussions of a Social Charter in the EU, however, the perfunctory debate about citizens' rights was quickly transformed into a discussion of workers' rights (Culpitt 1992, Meehan 1993), reinforcing a trend that already existed in many Member States towards a 'bifurcated welfare model' (Abrahamson 1992). This offers some basic protection for workers but little if any to the more marginalised populations in that a dual welfare system is fostered whereby company-based or employment-related welfare schemes take care of those in employment but neglect or exclude marginal and less privileged groups. Thus, and as Bennington *et al.* (1992) suggest, many were concerned that a corporatist model of welfare would evolve in the EU in which social rights are attached primarily to employment rather than to citizenship.

In the early years of the EU, little has occurred to dissipate these fears. The reality has been that the Social Charter has continued to focus upon workers' rights and thus exacerbated inequalities between those with and without employment. Even its attempts to introduce workers' rights, however, have met with only limited success. Pressures not only within but also external to the EU have limited progress on this matter. For instance, competition on wage costs from countries outside the EU such as those in South East Asia and Central and Eastern Europe, has put great pressure on the EU nations to keep down their social costs.

Indeed, the evidence now available suggests that such constraints have resulted in cutbacks in formal welfare provision. Between 1993 and 1998, expenditure on social protection as a percentage of GDP decreased (European Commission 2001b). This was most pronounced in those countries where

Table 3.1 Expenditure on social protection as a percentage of GDP in European Union nations, 1993 and 1998

	1993	*1998*	*% change*
EU-15	28.9	27.7	–1.2
Sweden	38.6	33.3	–5.3
Finland	34.6	27.2	–7.4
Netherlands	33.5	28.5	–5.0
Denmark	31.9	30.0	–1.9
France	30.9	30.5	–0.4
Belgium	29.5	27.5	–2.0
UK	29.1	26.8	–2.3
Austria	28.9	28.4	–0.5
Germany	28.4	29.3	+0.9
Italy	26.2	25.2	–1.0
Spain	24.7	21.6	–3.1
Luxembourg	24.5	24.1	–0.4
Greece	22.3	24.5	+2.2
Portugal	21.3	23.4	+2.1
Ireland	20.5	16.1	–4.4

Source: Eurostat – European System of Integrated Social Protection Statistics (ESSPROS). Cited in European Commission (2001b: 91).

spending had been amongst the highest in 1993 such as Sweden (–5.3 percentage points), Finland (–7.4 points) and the Netherlands (–5.0 points). As feared at the advent of the EU, therefore, there has been a levelling down of social protection. Those nations whose level of social protection has decreased most sharply are those who had the highest level of social protection in the first place (see Table 3.1). Meanwhile, and with the exceptions of Ireland and Spain, those with the lowest levels of social protection in 1993 have converged towards the norm. The result is that although there is a convergence of social protection between member states, this is happening within the context of an overall levelling down process.

This evidence of a levelling down should be no surprise. Even Keynes and Beveridge, the founders of the welfare state, recognised that the foundations of social welfare lie in the formal labour market. For them, full-employment, not welfare states, was the key to economic well-being. Full-employment meant low demand for social transfers and a large tax base to finance social programmes for the aged, sick and the minority of persons without jobs. Comprehensive formal welfare states were possible only so long as most people found their 'welfare' in the market most of the time (Myles 1996). The persistently wide 'jobs gap', not to mention an ageing population, in the advanced economies thus has profound implications for the future of comprehensive formal welfare provision.

If the goal of a full-employment and/or comprehensive formal welfare provision is beyond reach, the question then arises of whether it is appropriate to seek the eradication of work beyond employment. In this repressive discourse where a blind faith is put in a return to full-employment and comprehensive welfare provision, little consideration is given to the notion that this might not be achievable and that people need to be given alternative modes of work and welfare.

However, it is not only the problems involved in achieving full-employment and a comprehensive formal welfare state that lead one to be cautious about the idea of repressing the informal sphere. There is also the question of whether it is in fact possible and/or desirable to eliminate such activity. A major practical problem with seeking the eradication of informal work is that there are 'resistance cultures'. Many do not wish to reduce their informal economic activity and it will be difficult to persuade them to do otherwise. As will be shown in Part II of this book, such activity is deeply embedded in everyday social life, often undertaken for social as much as for economic reasons. Quite how and whether one can stop such 'social' activity is not immediately apparent. Nor arguably is the repression of such activity desirable, not least due to its impacts upon social cohesion, reciprocity, civil society and social support.

In sum, repressive discourses are founded upon a belief that we can return to a world of full-employment and/or comprehensive formal welfare provision. Once one accepts that this ideal-type is unlikely to be achieved, then a question mark arises over whether it is appropriate to repress informal work without putting in place any alternatives.

LAISSEZ-FAIRE DISCOURSES ON INFORMAL WORK

A laissez-faire approach towards informal work, or what might be alternatively called a de-regulationist discourse (DRED), is adopted in the main by neo-liberals. Similar to old-style social democrats, neo-liberals adopt the goal of full-employment and view employment as the best route out of poverty. However, the means by which this is to be achieved is very different. For neo-liberals, over-regulation of the market is to blame for many of the economic ills befalling society (Amado and Stoffaes 1980, Minc 1982, Sauvy 1984, de Soto 1989, Stoleru 1982). As Peck (1996: 1) summarises, 'From this viewpoint, failure is seen to have occurred in the market, not because of the market'. The solution, in consequence, is to liberate the labour market from 'external interference' so as to give market forces free reign (Minc 1980, 1982, Sauvy 1984, Stoleru 1982).

Included under the umbrella of external interference is state intervention. Antagonism to the welfare state, for example, is one of the prominent neo-liberal beliefs. The welfare state is seen as the source of all evils in much the same way as capitalism was by the left. What provides welfare if the welfare state is to be dismantled? The answer is market-led economic growth. Welfare here is understood not as state benefits but as maximising economic progress by allowing markets to work. Despite many assuming that neo-liberals differ markedly to old-style social democrats on the welfare issue, there are some common threads. Both view the welfare state and the economy as adversaries in that one is usually seen as the root cause of problems in the other. The difference is that whilst old-style social democrats favour the welfare state and view free market capitalism as causing poverty and inequality, neo-liberals support the free market and dislike any structure that constrains it. Old-style social democrats, therefore, view the welfare state as a necessary institution for the functioning of modern welfare capitalism and a prerequisite for efficiency and growth as well as individual self-realisation. Neo-liberals, in contrast, view the adversarial relationship between the welfare state and economic efficiency in the opposite manner. The welfare state is seen to interfere with individual freedoms and the ability of the market to optimise the efficient allocation of scarce resources.

Considering this view of the relationship between welfare and economy, a long-standing debate in neo-liberal thought has been the extent to which a welfare state is required. As Esping-Anderson (1994) displays, it has been with us ever since the English Poor Law reforms in the early part of the nineteenth century. Within the tradition of classical economics and libertarian thought, one extreme, exemplified by Smiles (1996), held that virtually any socially guaranteed means of livelihood to the able-bodied would pervert work incentives and individual mobility. This, in turn, would stifle the market, freedom, and prosperity. This is echoed today by those considering welfare provision to be the antithesis of social equality. As social rights are essentially claims against the income and resources of others, the welfare state is not considered a guarantor of equal status and autonomy. It is viewed as a divisive system under which a class of claimants becomes parasitic upon others' labour and property, with disastrous effects upon their morals (Gray 1997, Murray 1984). Others however, such as Adam Smith, realised that society did need social provision, especially in health and education, but stressed that due to the conflicting relationship between welfare and economy, this incurs a certain price in terms of economic performance. For these neo-liberals, therefore, the issue is where to set the trade-off between equality and efficiency (Barr 1992, Gilder 1981, Lindbeck 1981, Okun 1975).

There is thus a spectrum of neo-liberal thought that ranges from those who see no need for a welfare state at one pole to those who regard the need for lesser degrees of emphasis on efficiency as one moves along the continuum towards the other pole. All positions on the spectrum, however, emphasise the efficiency trade-offs of pursuing greater equality, particularly with reference to the possibly negative effect of the welfare state on savings (and hence investment), work incentives (and hence productivity and output) and the institutional rigidities that welfare states introduce (such as in the mobility of labour).

Although such internal debates over the degree to which a welfare state should be provided are important to adherents of this approach, the fundamental fact should not be masked that neo-liberals are on the whole negative about how the welfare state influences economic performance. For them, competitive self-regulatory markets are superior allocation mechanisms from the viewpoint of both efficiency and justice. Government interference in distribution (aside from marginal cases of imperfections, externalities or market failure) thus risks generating crowding-out effects, maldistribution and inefficiency and the end result will be that the economy will produce less aggregate wealth than if it were left alone (Lindbeck 1981, Okun 1975). Some even go so far as to insist that inequalities must be accepted, and perhaps even encouraged, because their combined disciplinary and motivational effects are the backbone of effort, efficiency and productivity (Gilder 1981).

This neo-liberal view would perhaps be harmless if it was simply an academic theorisation. However, it is not. Despite being a theory that is heavily opposed to state-led change, it is ironically the state in Anglo-Saxon nations, such as the UK and US and to a lesser extent Canada and Australia, which has been the primary vehicle for the implementation of this ideology. Indeed, this revelation is not new. Over fifty years ago, Polanyi (1944: 140) recognised that 'the road to the free market was opened and kept open by an enormous increase in continuous, centrally organised, and controlled interventionism'.

What is the approach towards informal work, therefore, in this neo-liberal view? On the whole, a positive perspective is adopted towards those who partake in such activity but there is largely seen to be no need to directly intervene in such activity in order to cultivate it. For these neo-liberals, such work is seen in two ways. First, it is seen as an indicator of how the formal sphere might be organised if it were de-regulated and second, it is viewed as a principal means used by the unemployed and marginalised as a means of getting-by (Matthews 1983, Minc 1982, Sauvy 1984, de Soto 1989). Indeed, the only change required is that the welfare 'safety net' needs to be dismantled (e.g. Matthews 1983) so as to allow civil society to freely operate as a self-generating mechanism of social solidarity. The little platoons of civil society must be allowed to flourish, and will do so if unhampered by state intervention. The virtues of civil society, if left to its own devices, are asserted to include 'Good character, honesty, duty, self-sacrifice, honour, service, self-discipline, toleration, respect, justice, self-improvement, trust, civility, fortitude, courage, integrity, diligence, patriotism, consideration for others, thrift and reverence' (Green 1993: viii). The state, it is believed, suppresses all these attributes and if they are to flourish, the state must withdraw from its welfare interventions.

The state, particularly the welfare state, is thus seen as destructive of the civil order, particularly obligations and duties associated with the family. Informal work, meanwhile, is the people's 'spontaneous and creative response to the state's incapacity to satisfy the basic needs of the impoverished masses' (de Soto 1989: xiv–xv). As Sauvy (1984: 274) explains, such work represents 'the oil in the wheels, the infinite adjustment mechanism' in the economy.

It is the elastic in the system that allows a snug fit of supply to demand that is the aim of every economy. If left to operate unhampered by the state, then the family and civil society would be able to provide this snug fit.

Critical evaluation of a laissez-faire approach

This approach is based on the idea that full-employment can and will return if market forces are allowed to operate unhindered. Measured purely in terms of whether this neo-liberal approach is more effective in achieving the goal of full-employment than the social democratic approach, there is little doubt that this is the case. It does indeed appear that unemployment is lower in advanced economies such as the UK and US that have pursued a neo-liberal strategy than it is in more social democratic nations of mainland Western Europe (see Chapter 1).

However, sound performance in terms of job creation must be seen in terms of both the quality of the jobs created and the degree of social polarisation that this 'success' has entailed (Conroy 1996, Esping-Anderson 1996, European Commission 1996b, Fainstein 1996, OECD 1993, Peck 1996). Whatever way social polarisation is assessed (see Pinch 1993, Williams and Windebank 1995a, Woodward 1995), the finding is that it is far greater in neo-liberal economies than social democratic nations. For example, and as Chapter 1 displays (see Table 1.2), the US and the UK, as the two nations that have led the race towards de-regulation, are also the nations with the highest levels of social polarisation between households. These nations have the highest proportion of households that have either multiple or no earners. We can thus only agree with Peck (1996: 2) that, 'Contrary to the nostrums of neo-liberal ideology and neo-classical economics, the hidden hand of the market is not an even hand.'

A further problem is that there is little evidence that some of the fundamental tenets of this ideology will have the impact neo-liberals desire. Take, for example, the policy of stripping away the welfare state so as to encourage people to find employment. Numerous studies of the effects of reducing welfare benefits on the levels of unemployment conclude that decreasing levels of benefit (or withholding benefit) will not cause an increase in flows off the unemployment register (Atkinson and Micklewright 1991, Dawes 1993, Deakin and Wilkinson 1991/2, Dilnot 1992, Evason and Woods 1995, McLaughlin 1994).

In an extensive review of the effect of benefits on (un)employment, McLaughlin (1994) concludes that the level of unemployment does have some impact on the duration of individuals' unemployment spells, but the effect is a rather small one. Following Atkinson and Micklewright (1991) and Dilnot (1992), she states that the level of unemployment benefits in the UK could not be said to contribute to an explanation of unemployment to a degree that is useful when considering policy. Moreover, extremely far-reaching cuts would be required in benefit levels to have any significant impact on the duration and level of unemployment. The effect of such cuts would be to create a regime so different from the present one from which the estimates of elasticities (of unemployment duration with respect to out-of-work benefits) are derived, that their predictive usefulness would be very suspect.

Neither will taking away the cushion of the welfare state simply allow the little platoons of civil society to spring into action and people to get by on their own. As displayed in Chapter 2, vast numbers of studies now detail how marginalised populations are less able to engage in informal work than more affluent populations. The result is that a laissez-faire approach towards informal work will merely consolidate, rather than reduce, socio-spatial in-equalities. Those least able to participate in informal work will remain unable to perform such activity as a coping practice. Therefore, the possibility that informal work can substitute for formal provision simply by taking away people's safety net appears mistaken. The result is more likely to be that people will be simply left bereft of the means of survival (see, for example, Mingione 1991).

Nevertheless, given the increasingly untenable pressures being put on formal welfare provision in the advanced economies, it remains obvious that something needs to be done about the way in which welfare is provided. However, the choice is not either to strive to restore full-employment and a comprehensive formal welfare 'safety net' as the old-style social democrats advocate or to strip away the welfare safety net and de-regulate the formal labour market as a means of encouraging full-employment, as the neo-liberals propose. As shown, the former is impractical and flies in the face of the direc-tion of the advanced economies, whilst the latter, even if it were to achieve full-employment, would do so at an extremely high price in terms of the levels of absolute and relative poverty. Given that the repressive and laissez-faire approaches towards informal work leave marginalised populations unable to engage in alternative means of livelihood and welfare provision, another approach towards informal work is here considered: that of enabling people to engage in such activity.

CULTIVATING THE INFORMAL SPHERE AS A SPRINGBOARD INTO EMPLOYMENT

To distinguish the contrasting approaches towards cultivating informal work, two key questions need to be asked. What type of informal work are they seeking to harness? And for what purpose is the informal sphere being culti-vated? The first approach considered here is that which seeks to harness various types of informal work as a springboard into employment. Similar to previous sections, we here first outline the approach and second, evaluate critically its implications.

The approach that seeks to cultivate informal work as a springboard into employment is no different so far as the future of work is concerned from the above discourses. It seeks to achieve full-employment. The difference between this and previous approaches, however, is that the 'social economy' or 'third sector' is seen as an additional means of job creation and/or improving employability, as is the formalisation of micro-level informal work (e.g. childcare).

The rationale is that in an age of de-coupling of productivity increases from employment growth, the private sector can no longer be relied upon to create sufficient jobs. Neither, moreover, can the post-war corporatist welfare state model be expected to spend its way out of economic problems. The 'fiscal crisis' of the state, widely predicted in the 1970s (Habermas 1975, O'Connor 1973), is now seen as firmly established. The new realities of the global economy are seen to have forced states radically to reorient their regeneration strategies (Amin *et al.* 1999, Peck 1998). The 'social economy' is thus bolted on to conventional job creation and training programmes and policies.

This approach has been widely adopted. As evidence of its prevalence, for example, one has only to note that the European Commission's major mechanism to stimulate the third sector is entitled the 'Third System and Employment' (see ECOTEC 1998, Haughton 1998, Westerdahl and Westlund 1998). Arguably, this is also the approach of the New Labour government in the UK. Through its urban and neighbourhood regeneration policies masterminded by the Social Exclusion Unit (SEU) and the Neighbourhood Renewal Unit (NRU), the UK government has shown itself to be genuinely supportive of third sector initiatives and community enterprises. This, however, is so only in the context of markedly deprived areas and as a stepping-stone back into the formal economy (Amin *et al.* 2002). The third sector is something to be cultivated to aid those excluded from the formal labour market to be either directly re-included into the formal labour market or indirectly by improving their employability.

To achieve this, it is the more formal community-based organisations that are focused upon. Viewing informal work in terms of a 'ladder' or 'hierarchy' of types (see Home Office 1999), at the top of this ladder are 'third sector' and/or relatively formal community-based groups with micro-level informal forms of one-to-one reciprocity at the bottom. Only the 'mature' forms of work are viewed as worthy of cultivation as a new source of formal work and means of improving employability. The lower echelons of informal work are viewed, in contrast, as ripe for formalisation (e.g. parental work, caring). In this approach, no explicit distinction is drawn between these two contrasting strategies towards informal work. Both are ways of creating formal work and/or improving employability. Integration in this approach, therefore, means integration into the mainstream of formal work and the informal sphere is a tool to be harnessed to achieve this end. Informal modes of production are simply an additional means by which the 'jobs gap' left by the public and private sectors can be filled (Archibugi 2000, Community Development Foundation 1995, ECOTEC 1998, European Commission 1996b, 1997, 1998b, Fordham 1995, OECD 1996).

Critical evaluation of an 'employment-focused' approach towards harnessing the informal sphere

In this approach, in consequence, the informal sphere is cultivated in order to achieve full-employment. This raises several issues. First, Giddens (1998) argues that we should beware of two forms of misplaced nostalgia. Left nostalgia is for the golden age of the developed welfare state coupled to full-employment while Right nostalgia is a longing for a traditional family that has been somehow corrupted by the state.

Examining this particular approach, there is little doubt that it has not yet managed to transcend the traditional nostalgia of the Left for full-employment. Its policy orientation is to use the informal realm to engender such a scenario. The key question, however, is whether harnessing the informal sphere can help to achieve this goal. As outlined in Chapter 1, the 'jobs gap' that needs to be bridged is vast. Whether the third sector or formal-ising informal work (e.g. childcare) can achieve this is thus open to question.

On the one hand, there is now a great deal of evidence that the third sector or social economy is unable to create formal jobs in large enough numbers to compensate for the inability of the private and public sectors to provide sufficient opportunities (e.g. Amin *et al.* 2002). On the other hand, there is mounting evidence that government attempts to formalise informal work are met with strong cultures of resistance that prevent large numbers of formal jobs being created. One evaluation of a Danish government initiative to create a household service market called Home Service, for example, found that besides 'economic' or cost considerations, strong social norms prevailed that led households to want to do tasks themselves (Sundbo 1997). This prevented the formalisation of much informal work. In Part II, this finding will be further reinforced where strong cultures of resistance to the formal-isation of informal work are identified amongst the English population. Just because governments might wish to harness the third sector and formalise micro-level informal work in order to create jobs does not mean that they will be successful.

The second problem with this approach of cultivating informal work as a means of jobs creation and improving employability is, as Amin *et al.* (1999, 2002) and Levitas (1998) point out, that it solely seems to be applied to marginalised populations at present. The consequence is that it carries the inherent danger that while the majority in employment participate in the private and public sectors, those marginalised from employment will find themselves increasingly inserted into what might be perceived as a 'second class and second rate' sphere of employment. To avoid this, then the employ-ment created will need to be of equal status in terms of income and rights as the jobs found in the public and private sectors and/or such employment will need to be targeted at everybody rather than just the poor. If this is not achieved, then the inevitable end result will be a dual society. A second-class form of employment will arise for those who would otherwise be margin-alised from the formal sphere.

A third issue with this approach is the way in which it differently views various forms of informal work. While there is a desire to cultivate more organised community-based groups as sources of employment, micro-level informal work (e.g. childcare by parents) is viewed as ripe for formalisation. 'New Labour's position assumes an individual can contribute to society through working as a paid child carer but not as an unpaid mother, though both individuals carry out the same tasks and make identical contributions' (King and Wickham-Jones 1998: 277). This has been argued by many other critics (e.g. Hills 1998, Levitas 1998, Lister 1997). Voluntary work, meanwhile, is seen as a way of improving employability by cultivating skills, self-confidence, self-esteem and the ability to work with others (Levitas 1998). Why these two forms of micro-level informal work should be differently viewed is difficult to discern. Indeed, there is a very fine line between the eradication approach of the Old Left and this approach that seeks to 'harness' work beyond employment.

Fourth and finally, and on the issue of cultivating such activity as a springboard into employment, the lesson from a wide range of studies is that this is not feasible (e.g. Amin *et al.* 2002, Williams *et al.* 2001). Many studies display that the social economy can play only a very minor role in inserting people into jobs. This has been explicitly discussed in the UK, for example, in the case of Local Exchange and Trading Schemes (LETS). This displays how LETS is ineffective as vehicles for creating jobs and inserting people into the formal labour market, although they are effective at providing members with alternative means of livelihood (see Chapter 8). To use such initiatives as springboards into employment, therefore, is to ignore what they are most effective at doing. In sum, in this 'mono-economic' approach, informal work is harnessed solely as a means of providing formal work. It retains a steadfast belief that formal employment represents the route out of poverty.

HARNESSING INFORMAL WORK AS A COMPLEMENT TO FORMAL WORK AND/OR WELFARE PROVISION

Rather than harness informal work in order to provide an additional springboard into employment and means of improving employability, another approach is to view informal work as a complement to formal employment and/or welfare provision. Here, informal work is seen as an additional economic tool providing means of livelihood that can supplement formal employment and/or as a welfare tool that can supplement private and public welfare provision. As will be argued below, there is a need to distinguish two distinct sub-approaches here. On the one hand, there are those who solely wish to harness the informal sphere as a supplementary means of welfare provision (e.g. Giddens 1998, 2000, 2002). On the other hand, there are those who view it also as an economic tool that can provide a complementary means of livelihood to supplement the formal labour market (e.g. Amin *et al.* 2002, Aznar 1981, Beck 2000, Delors 1979, Greffe 1981, Jordan 1998, Laville 1995, 1996, Mayo 1996, Roustang 1987, Sachs 1984, Williams and Windebank 1999b, 2001b, 2001d).

Informal sphere as a complement to private and public welfare provision

Let us start with those who view the informal sphere as a welfare mechanism that can complement public and private welfare provision (rather than as an economic tool). In this discourse, informal provision is a means of filling the 'welfare gap' left by the public and private spheres. By cultivating the informal sphere, it is believed that many welfare needs that are currently unmet by the formal public and private spheres can be fulfilled by community-based groups, voluntary work and one-to-one reciprocity.

In recent years, there has been a good deal of confusion and debate over the approach that New Labour adopts towards work beyond employment (see, for example, Jordan 1998, Jordan and Jordan 2000, Levitas 1998). Our view is that in order to understand the approach of New Labour, it is first necessary to recognise that a clear (if artificial) distinction is drawn between the role attached to informal work in the 'economic' sphere and its contribution in the realm of 'welfare'. In the 'economic' realm, New Labour views work beyond employment purely as something to be cultivated as a means of achieving job creation and improving employability. In the 'social' or 'welfare' sphere, however, New Labour recognises informal work in its own right as a form of activity that can complement the private and public formal spheres as an additional means of welfare provision.

In the realm of welfare, in consequence, informal provision or what is sometimes referred to as civil society, is viewed as a third prong in the welfare equation, which can act alongside public and private sector provision in meeting welfare needs. As Giddens (2000: 55–6) puts it,

> The 'design options' offered by the two rival political positions were ministic – they looked either to government or to the market as the means of co-ordinating the social realm. Others have turned to the community or civil society as the ultimate sources of social cohesion. However, social order, democracy and social justice cannot be developed where one of these sets of institutions is dominant. A balance between them is required for a pluralistic society to be sustained.

In the third way approach of New Labour, therefore, the informal sphere is harnessed so as to provide a mixed economy of welfare. Transcending the public versus private provision debate, civil society is added into the equation as an additional means of welfare provision. As Giddens (2000: 81–2) continues,

In the past, some on the left have viewed the 'third sector' (the voluntary sector) with suspicion. Government and other professional agencies should as far as possible take over from third-sector groups, which are often amateurish and dependent upon erratic charitable impulses. Developed in an effective manner, however, third-sector groups can offer choice and responsiveness in the delivery of public services. They can also help promote local civic culture and forms of community development.

Or as the UK Prime Minister, Tony Blair (1998: 14) puts it,

The Old Left sometimes claimed that the state should largely subsume civil society, the New Right believes that if the state retreats from social duties, civic activism will automatically fill the void. The Third Way recognises the limits of government in the social sphere, but also the need for government, within those limits, to forge new partnerships with the voluntary sector . . . 'enabling' government strengthens civil society rather than weakens it, and helps families and communities improve their own performance.

If New Labour accepts that the development of the informal realm provides a third prong in the welfare equation that complements private and public provision, it is necessary to reiterate, however, that this is not the case when viewed as an 'economic' tool. Here, their approach remains entrenched in an ideology of formalisation and there is a fine line between their rhetoric of cultivating the third sector as a means of job creation and improving employability, and the old-style social democratic approach of repressing informal economic activity.

Informal 'economy' as a complement to formal employment

For others, however, informal work represents a complement not just to the private and public sectors in the welfare sphere but also in the 'economic' realm (e.g. Amin *et al.* 2002, Aznar 1981, Beck 2000, Delors 1979, Greffe 1981, Jordan 1998, Laville 1995, 1996, Mayo 1996, Roustang 1987, Sachs 1984, Williams and Windebank 1999b, 2001b). Here, inserting people into employment is not the only (or even the most effective) way to ensure that people are able to meet their needs and creative desires. Although employment might be a 'route out of poverty', it is certainly not the only, or the best, route for these analysts. If needs are to be met and creative desires fulfilled, inserting people into formal employment so as to earn money to pay for formal goods and services that meet needs or desires is one option. Another path is to help people receive these same goods and services through informal modes of production.

Here, therefore, getting a formal job in order to earn money so as to pay somebody else on a formal basis to provide a good or service is not the only way of resolving poverty. There are also more direct routes. A new balance between employment on the one hand, and non-market activities on the other, is thus sought in the form of a 'mixed economy'. However, this mixed or plural economy is not advocated solely in the realm of welfare provision as in the above approach that calls for 'mixed economies of welfare'. Instead, it is also applied to the organisation of working time. For these analysts, there is also a need to develop a more pluralistic mixed economy in the form of the public (state), private (market) and informal (non-market work) economic spheres that it believes can live harmoniously together.

In this rethinking of what constitutes 'the economic', the argument about how this is to be achieved differs on three counts from the laissez-faire approach advocated above that also seeks to promote informal work. First, it is based upon the logic of supplementing and not substituting formal activity and state provision. Second, it is based on the concept of optionality and choice, which contradicts the conservative appeal to duties and norms. Third and finally, such informal work is envisaged more in terms of collective and interactive forms of working instead of in terms of isolation and competition (see Evers and Wintersberger 1988).

Rather than pursue the precarious strategy of putting all of one's eggs into the one policy basket of job creation, this approach thus seeks to develop complementary means of livelihood (e.g. Beck 2000, Williams and Windebank 1999b, 2001b). The reasons for this approach of pursuing 'active policies' in relation to not only insertion into employment but also informal work are due to first, the current inequalities in the ability of people to participate in these other subsystems of work and second, the niggling doubt in their minds over whether full-employment can be (and should be) achieved and maintained.

As displayed in Chapter 2, informal work is currently unable to act as a substitute for those excluded from employment. This is widely recognised by these analysts. In consequence, 'complementary social inclusion policies' (CSIPs) are advocated to harness informal work as additional means of livelihood in order to open up these activities to those currently excluded from engagement in such work (Williams and Windebank 1999b). Just as intervention is required in the formal sphere to prevent the socio-spatial inequalities that arise when the market is left to operate unhindered, so, as this approach asserts, similar interventions are required in the realm of informal work. Unless active policies are adopted, the current socio-spatial inequalities in informal work (identified in Chapter 2) will persist and the poor will continue to be unable to mitigate their circumstances through their informal economic activity.

A second and perhaps more important reason why active policies are advocated relates to the doubt that gnaws away in the minds of most of these analysts that perhaps full-employment cannot (and even should not) be achieved. Cutting through the party political hype that the achievement of full-employment is nearly upon us, these analysts argue that such a situation is far from the reality (e.g. Beck 2000, Rifkin 1995, Williams and Windebank 2001b). To do so, they use very similar arguments to those propounded in Chapter 1. Pointing out that there is a large gap between current employment participation rates and a full-employment scenario, their concern is what is to be done with the large segment of the population who are likely to remain excluded from the formal sphere. For them, the informal sphere is one solution.

For this approach, moreover, it is not just the feasibility of creating sufficient jobs in order to achieve full-employment that is important. The concern is, to repeat, whether one should put all of one's eggs in the policy basket of creating sufficient jobs. For them, it seems logical also to consider developing alternative means of livelihood in case this uni-dimensional and 'mono-economic' policy approach does not succeed. Indeed, it is not only the lack of employment opportunities for all that leads to this conclusion. It is also as outlined in Chapter 1 the desirability of pursuing full-employment (e.g. Beck 2000, Gorz 1999, Rifkin 1995, Williams and Windebank 2001b).

In consequence, the intention in this approach is not to 'colonise' (Heinze and Olk 1982) the informal sphere by incorporating it into the formal wage system. Neither is it to develop the informal sphere as an alternative to formal employment (see below). Rather, the intention is both to pursue the promotion of formal employment and, at the same time, harness informal work so as to enable the development of complementary means of livelihood. This approach towards the informal sphere has been given various names: 'liberal self-help' (Evers and Wintersberger 1988), a 'complementary' approach to local economic development (Macfarlane 1996), a 'complementary network' model (Heinze and Olk 1982), 'DIY citizenship' (Simey 1996) and a 'social economy' approach (Porritt 1996). Whatever the name given, the important point is that it aims to use the informal sphere to 'level the playing field' and empower disenfranchised individuals and communities through capacity-building to help themselves.

A critical evaluation of the complementary approach

Whether analysts wish to use the informal sphere solely as a complementary welfare tool or also as an 'economic' tool, a common problem is confronted. The danger is that this 'dual economy' approach will result in the creation of a dual society. In theory, this is not inevitable. In practice, however, the current calls end up advocating such an outcome, often unintentionally. Take, for example, the advocacy of civil society as a complement to public and private welfare provision. On the whole, it is only the poor and unemployed to which this is applied. In the UK for instance, the Social Exclusion Unit and Neighbourhood Renewal Unit apply this largely to deprived populations (e.g. Social Exclusion Unit 1998, 2000). It is not more widely advocated to the population at large who instead are encouraged to seek their welfare in the formal sphere. If pursued, therefore, such activity will become a welfare tool only for those unable to access private and public formal welfare provision. The outcome will be the advent of a dual society.

Such (unintentional) advocacy of a dual society is not confined to the policy-making community. Academics who support the development of informal provision as a complementary welfare tool also end up unintentionally doing so. As Giddens (1998: 110) argues,

> *Conventional poverty programmes* need to be replaced with community-focused approaches, which permit more democratic participation as well as being more effective. Community building emphasises support networks, self-help and the cultivation of social capital as means to generate economic renewal in *low-income neighbourhoods* [our emphasis]

Such complementary mechanisms are not advocated more widely. For more affluent communities, the perception is that private (or public) sector formal provision is the way forward. This is not to say that these analysts are not conscious of the dangers of creating a dual society. Indeed, many actively seek not to do so. However, their discussion often results in an analysis of the role of work beyond employment in tackling the problems of marginalised groups. This is due to both their pragmatic attempts to show how such activities can be a useful complement to the formal sphere and their desire to solve the dire problems of these groups. If such a focus is then translated into policy, however, the end result will be the creation of a dual society.

One way forward has been to recognise that informal modes of work and welfare provision should not be applied solely to the poor and socially excluded but to everybody. Making a concerted effort to cultivate the informal sphere amongst not only the poor and socially excluded but all the population, however, could well reinforce, rather than reduce, socio-spatial inequalities. It is now well understood that affluent populations are more able to draw upon the resources of the informal sphere than disadvantaged populations (see Chapter 2). If the informal sphere is not to reinforce socio-spatial inequalities, complementary social inclusion policies will thus need to continue to focus upon harnessing such activity amongst deprived populations, leading to the accusation that a second-class second rate realm of provision for the poor and marginalised is being created. It is thus difficult to see how to avoid accusations that a dual society is being created.

For some, the only solution to this dilemma, adopted mostly by those who wish to pursue a post-capitalist more associative economy, is to drop the notion that the informal sphere should be seen as a complement to formal provision. For them, the whole point of developing the informal sphere is to create an alternative to capitalism and they argue that it is little use trying to be pragmatic and advocating it as an additional tool for complementing formal provision. To do so only results in the co-option of the informal sphere by capitalism as a means of offsetting its costs and offloading those no longer of use in terms of profit-motivated market exchange. Such a 'dual society' approach is thus an inevitable consequence of viewing the informal sphere as a complement to employment.

HARNESSING THE INFORMAL SPHERE AS AN ALTERNATIVE TO FORMAL WORK AND WELFARE

The final discourse on harnessing the informal sphere thus views it as an alternative to formal work and welfare provision. This approach is grounded in the two contrasting political traditions briefly introduced in the introduction, namely communitarian ecocentrism and/or the radical European social democratic approach. Here, we review in a little more depth each of these traditions in turn so as to understand the ideological roots of this approach.

Radical European social democrats

The first stream of thought that advocates the development of alternatives to formal work and welfare is what we here call the radical European social democrat tradition (e.g. Aznar 1981, Beck 2000, Delors 1979, Gorz 1999, Lalonde and Simmonet 1978, Sachs 1984). For these radicals, advocating the development of such work is embedded in a broader canvas of analysis, critique and prescription. To understand their approach, we outline first how they define present-day problems of work and second, their analysis of the role informal work can play in resolving these problems.

For most of these analysts, there are three interrelated problems surrounding the current organisation of work. First, it is maintained that most people are unsatisfied with their formal employment because their jobs are often stultifying, alienating and do not allow them the freedom to compensate for this lack of satisfaction outside working hours (e.g. Archibugi 2000, Aznar 1981, Gorz 1985, 1999, Laville 1995, 1996, Mayo 1996). As Archibugi (2000: 9) puts it, employees 'are all deeply isolated and dissatisfied. The workplace and the activity no longer seem to be "places" of social integration.' Given this lack of opportunity for personal growth in employment as well as the fact that the only alternative to a job is unemployment that cannot provide self-esteem, social respect, self-identity, companionship and time structure, the age of employment is not perceived as at all positive. Aznar (1981: 39) assesses this situation in the following manner:

> any society which proposes that its citizens spend the whole of their time, energy and empathy engaged in an activity which cannot, by its very nature, soak up this energy, is fundamentally perverse.

The solution, as Beck (2000: 58) puts it, is that

> the idea that social identity and status depend only upon a person's occupation and career must be taken apart and abandoned, so that social esteem and security are really uncoupled from paid employment.

This is reinforced by Gorz (1999: 72) who argues that,

> The imperative need for a sufficient, regular income is one thing. The need to act, to strive, to test oneself against others and be appreciated by them is quite another. Capitalism systematically links the two, conflates them, and upon that conflation establishes capital's power and its ideological hold on people's minds.

The second reason why employment should not remain centre stage flows from their observation that society, in pursuing economic growth for its own sake rather than as a means to an end, has lost its way (Douthwaite 1996, Gorz 1985, Mayo 1996, Robertson 1985). In consequence, an individual's work, as embodied in employment, no longer responds to the real needs of the consumer. The result, as Friedmann (1982) states, is that only 35–40 per cent of the economically active population engage in 'indispensable' production, the rest are obliged to produce essentially unnecessary goods in order to earn the income necessary for personal survival. Gorz (1985: 58) agrees and suggests that 'for a section of the population, only the production of inessentials allows them access to necessities'. In this discourse, however, nobody should have to work at such production to earn the money necessary for survival: 'employment should be seen not as an end in itself, but as a means to achieving a better quality of life' (Mayo 1996: 147).

Third and finally, these analysts argue that the relentless pursuit of full-employment has led to a devaluation of informal economic activity. The prolonged structural crisis of unemployment, however, is considered to provide an opportunity to valorise this form of work. As the OECD (1996) report produced by some advocates of this discourse suggests, there is a need to put the economy back into society rather than see it as an autonomous or independent element. For them, and mirroring much of the recent academic interest shown in analysts such as Polanyi (1944) and Granovetter (1973), the desire is to develop a socially embedded view of economic activity (see, for example, Lee 1996, Verschave 1996). The current structural crisis of unemployment for them provides just such an impetus for rethinking whether all social goals should be subjugated to the economic aim of continued growth rather than a view of the 'economy' as serving the interests of society.

Based on this definition of the crisis of work, they believe that informal work has a crucial role to play. These theorists are concerned not only, or indeed, necessarily, with informal activities as they exist today, but with the possible emergence or reinforcement of a category of work that one could call 'autonomous'. Such 'autonomous' work does not refer to a tangible empirically observable category of activity. Instead, the notion is linked to utopian visions of what work could be in the future. Thus, autonomous work is a conceptual, as opposed to a concrete, phenomenon. It suggests that a type of work should come about in the future over which the producer will have a large degree of control in which creativity and conviviality will be the driving forces. In sum, autonomous work is creative, controllable and socially useful. It has a purpose for the person performing it other than the earning of a wage.

On the whole, the most 'reformist' of these radicals have a vision of autonomous work that merges with present forms of informal work. In so doing, like Unger (1987), they view alternatives as emerging out of critical and practical engagements with the associations, personal behaviours and practices that now exist. The more 'transformative' theorists, however, do not explicitly refer to types of informal work that exist today since this does not fit their aim of calling for a complete redrawing of the boundaries that divide various categories of work from one another. Instead, and as Windebank (1991) asserts following her in-depth interviews with some of these theorists in France, the attempt is to find new forms of autonomy and sociability in the allocation of work. What unites all these radicals is their wish to put an end to, or at least considerably reduce, the domination of 'heteronomous' work, which is understood as that over which individuals have little or no control, and introduce more autonomous work.

There are two separate but connected future scenarios for work envisaged by these theorists. First, there is the 'socialist' scenario in which the state would take control of heteronomous production and designate the boundaries between the autonomous and heteronomous spheres of life. Second, there is a scenario that rests on the more libertarian and ecologist ideal of changes in work patterns coming about through piecemeal initiatives. Accordingly, the autonomous sphere would develop as a result of civil society nibbling away at the domains of the formal market and state.

In the first scenario, the desire is to bring about a self-managed or *auto-gestionnaire* society in which the informal sphere would play a major part. What renders this a 'socialist' scenario is that the state would retain an important role in organising the essential heteronomous work on which a society as a whole relies for the production of its staple goods (ADRET 1977, Gorz 1985, 1999). Gorz (1985), for example, calls for a new balance between three types of work: heteronomous work, small-scale co-operative and communal free enterprise, and autonomous household-based activity. The ADRET collective (1977), meanwhile, asserts that a pattern of organisation for work should evolve whereby each person would perform an average of two hours a day of 'constrained' work (which could be translated into so many days a week, months a year, and so on, according to individual tastes and circumstances). In conjunction with such work, there would be an absolute ceiling placed on the amount of goods and services produced in this manner. Such a new organisation of work could not come about, they argue, without a change in the way society and economy are organised. It cannot emerge within a productivist or growth economy fuelled by the capitalist profit motive. Thus, the role of the market economy is in some doubt in this scenario, being replaced for the most part, by the state, which it is argued, could organise heteronomous work.

The second scenario overlaps more with the communitarian ecocentrist perspective discussed below. It rests on the libertarian and ecologist ideal of changes in work patterns emerging through piecemeal initiatives. Accordingly, the autonomous sphere would develop as a result of civil society providing on a case-by-case and area-by-area basis alternatives to the formal market and state (e.g. Lalonde and Simmonet 1978, Lebreton 1978). These theorists advocate the encouragement of autonomous work not only as a means of improving people's quality of life, but because such diversified and decentralised activity is much more in harmony with the ecological principles by which they set so much store. These are dispersion, diffusion, diversity and dilution, which they contrast with the principles of concentration and simplification that dominate present economic matters.

Thus, they posit that growing autonomy on the part of individuals to work according to their own needs and desires (instead of according to those of industry) will result in a more diversified society and economy that will be more ecologically viable. Self-managed or autonomous work, therefore, is essential to environmental protection since it would allow society to become less concentrated and less ossified, more diverse and flexible and, as such, in keeping with and more responsive to the needs of the biosphere. For these theorists, in consequence, encouraging autonomous work is necessary not only to provide a more fulfilling life for individuals but also for ecological reasons.

The way forward is to develop the autonomous sphere by developing civil society initiatives that replace those currently conducted by the market and the state. Here, therefore, work beyond employment, at least to the extent that this is autonomous work, is seen as an alternative to employment and/or heteronomous work. It is, moreover, an alternative to employment for everybody, not just the socially excluded as is sometimes argued by those who view such activities as a complement to employment. Given that the view of informal work in this particular brand of radical European social democratic thought is very much akin to that of the radical ecologists, we here turn our attentions to this second body of thought.

Communitarian ecocentrism

For communitarian ecocentrists, alternatives to employment are sought more for rationales associated with the pursuit of ecologically sustainable development (e.g. Dobson 1993, Ekins and Max-Neef 1992, Fodor 1999,

Goldsmith *et al.* 1995, Henderson 1999, Hoogendijk 1993, Mander and Goldsmith 1996, McBurney 1990, Robertson 1998, Roseland 1998, Trainer 1996, Warburton 1998, Wright 1997). One of the starting premises of these communitarian ecocentrists is that socialism has become a spent political force due to both the changes in the political economy of Central and Eastern Europe and the parallel shifts that have been witnessed in the politics of First World 'advanced' economies. For them, environmentalism has replaced socialism as the new antithesis to capitalism. Unlike 'shallow' environmentalists who assert that environmental concerns can be accommodated within capitalism (see O'Riordan 1996), they argue that this is not possible and it is within this group that one finds a desire to develop alternatives to employment in particular and productivism and materialism more generally.

These ecologists do not believe that their views can be placed on the conventional left-to-right political continuum (e.g. Capra and Spretnak 1985, Dobson 1993, Eckersley 1992, Goodin 1992). This left-to-right continuum characterises the varying beliefs about how greater materialism can be realised for the majority of people and the differing stances about how to boost productivism. This perspective, however, agrees with neither of these objectives and thus places itself outside this left-to-right spectrum of political thought.

Rather than debate the best way to achieve materialism and productivism (the essence of the difference between old-style social democracy and neo-liberal thought), their argument is that no habitable future is possible without both a fundamental alteration of our attitude towards nature and a radical shift in the direction of society (Devall 1990, Devall and Sessions 1985, Eckersley 1992, Naess 1986, 1989, Skolimowski 1981). Here, therefore, we scrutinise this approach first, in terms of how it views the relationship between people and nature and second, in terms of how it believes that society's direction must alter. This will uncover the ways in which alternatives to employment are seen to play a central normative role in creating a more sustainable future.

So far as its view of the relationship between people and nature is concerned, its objective is to protect natural ecosystems but not simply for the pleasure of people as is so often the case in anthropocentrism. Instead, they argue that nature has biotic rights (e.g. Devall 1990, Naess 1989). Nature is seen to have a right to remain unmolested that does not require justification in human terms. The problem for these ecocentrists is that these biotic rights are not currently being respected. Writers such as Zimmerman (1987: 22) thus call for 'the elimination of the anthropocentric world view that portrays humanity itself as the source of all value and that depicts nature solely as raw material for human purposes'. Here, in consequence, anthropocentrism is replaced by biocentric egalitarianism, by which is meant inter-species equity that recognises non-human or biotic rights.

Grounded in this ecocentric premise, this approach then proceeds to argue that the human world in its endless pursuit of materialism is heading in the wrong direction. Their argument is that what were originally means to an end have become ends in themselves. For example, the acquisition of material goods was originally a means to achieving the end of well-being. Today, they argue, such a means has become an end in itself (e.g. Dobson 1993, Mander and Goldsmith 1996, Robertson 1985). They thus call for radical change. For them, there is a need to redefine 'wealth' as 'well-being' rather than the acquisition of material goods. In so doing, they mirror Aristotle's critique of moneymaking (chrematistics). Aristotle viewed the point of economic activity as enabling people to live well, which involved having sufficient time to spend developing friendships and the arts and participating in political deliberation. For this, production was directed towards use-values, and it was necessary to have some idea of sufficiency in material consumption. He saw economic activity directed towards moneymaking as pathological, as mistaking the means of achieving economic well-being for the ends (see Sayer 2001). Aristotle, of course, could not have envisaged that what for him was an aberration was to become a system imperative (Booth 1994).

In order to achieve such 'well-being' for both humans and non-humans, communitarian ecocentrists assert that a more small-scale decentralised way of life based upon greater self-reliance is required so as to create a social and economic system less destructive towards nature (e.g. Douthwaite 1996, Ekins and Max-Neef 1992, Gass 1996, Goldsmith *et al.* 1995, Henderson 1999, Lipietz 1995, Mander and Goldsmith 1996, McBurney 1990, Morehouse 1997, Robertson 1985, Roseland 1998, Trainer 1996). Progress means changing the way we live so as to achieve a more harmonious relationship with nature. To achieve this, the now established concept of 'thinking globally and acting locally' is the key. Global problems such as the destruction of nature can only be overcome by acting in a local manner (e.g. Mander and Goldsmith 1996). Take, for example, their view of the economy and economic progress. Rather than pursue the end of economic growth through outward-looking strategies, their objective is to ensure that the basic needs of all are met through the pursuit of self-reliance and an inward-looking approach (e.g. Ekins and Max-Neef 1992, Morehouse 1997, Robertson 1981, 1985).

For them, employment has a tendency to promote open economies. Informal economic activities, meanwhile, are seen to be more in keeping with their desire for inward-looking strategies and their objective of sustainable development (e.g. Henderson 1999, Mander and Goldsmith 1996, Warburton 1998). Therefore, the development of informal work resonates with their desire for a more localised self-reliant form of economic development. As Robertson (1981) puts it, developing informal work is essential to the creation of a 'saner, more humane and ecological' (SHE) future rather than a 'hyper-expansionist' (HE) future. For communitarian ecocentrists in other words, the development of informal work is a means of bringing about more self-reliant, sustainable economies.

This is often interpreted outside of this approach as a desire to return to some pre-industrial past based on self-sufficiency. It is important to state, however, that these ecologists do not seek for everything to be produced locally, nor do they seek an end to trade. They simply seek to forge a better balance between local, regional, national and international markets (Douthwaite 1996, Porritt 1996). They also seek to gain greater control over what is produced, where, when and how, so localities are less dependent upon the foibles of the global economic system for their future well-being. In other words, they seek self-reliance, not self-sufficiency. Far from reducing living standards, moreover, it is argued to make economic sense for a locality to seek to increase its net income and thus wealth, environmental sense to reduce unnecessary degradation and resource consumption and social and political sense to consider more directly meeting the needs and wants of citizens (see Williams 2000). For them, the promotion of informal work is central and essential to achieving such goals. There has thus been a strong tradition in this approach to advocate the development of informal work.

Evaluating informal work as an alternative to employment

For these analysts, a return to the supposedly 'golden age' of full-employment is unrealistic, illogical and undesirable (see, for example, Beck 2000, Gorz 1999, Macfarlane 1996, Williams and Windebank 1999a, 2001b). It is unrealistic because of the immense jobs gap that needs to be bridged and the marked lack of success in narrowing this gap despite all of the efforts to do so in the advanced economies. It is illogical because to seek a return to an age of full-employment presupposes that such an era actually existed. It is undesirable because of the degrading effects on people and nature of such a mode of economic organisation.

Therefore, rather than view informal provision as part of a strategy to achieve 'full-employment' and/or means of bolstering the sagging edifices of capitalism, this perspective instead interprets the principal role of this sphere to be one of facilitating a post-capitalist more associative society. For many, this will seem an unrealistic goal. Since the collapse of socialism, a wide-spread opinion has pervaded discourses in the advanced economies that 'there is no alternative' to capitalism. The only debate, as displayed so clearly in the discourses over globalisation, seems to be what type of capitalism is least and most desirable and how much intervention is possible. For this reason, few now look for alternatives to employment. The sole room for manoeuvre is seen to be the extent to which profit-motivated market exchange can be humanised. The value of the approach discussed here, however, is that it raises the spectre of an alternative. Resonating with most of what Giddens (2002: 11) identifies as the 'big changes' taking place – globalisation, the rise of individualism and post-materialist values, the dysfunctions of the welfare state and the emergence of new risks – this approach seeks to forge a new alternative to capitalism.

Nevertheless, and as Giddens (2002: 11) puts it, 'to pretend or imply that there is a known alternative to the market economy is a delusion'. We fully agree. This is not to say, however, that there is no alternative. In these approaches, even if not fully worked through, we find the seeds of an alternative future of work and welfare beyond capitalism that is grounded in harnessing non-market work and welfare and is focused upon developing people's capabilities to meet their needs and desires. The rest of this book is dedicated to making some attempt to further develop this approach.

CONCLUSIONS

The aim of this chapter has been to set out the various discourses on the informal sphere and their implications for policies towards such activity. This has revealed that the first option of more fully formalising work and welfare is both impractical and undesirable. It is impractical because informal work is deeply embedded in everyday life and the evidence points towards an informalisation rather than formalisation of work and welfare. It is undesirable because this work is often the preferred means by which people conduct many activities and a key ingredient of the social cement that binds communities together.

A laissez-faire approach, meanwhile, results in numerous negative consequences. The ever popular 'marginality thesis' assumes that informal work is undertaken by those marginalised from employment as a survival strategy for unadulterated economic reasons and is thus more prevalent in deprived than affluent communities. The empirical evidence of the vast majority of studies conducted throughout North America and Europe, however, is that informal work is not limited to the unemployed. Rather, the unemployed conduct less informal economic activity than the employed. As such, a laissez-faire approach merely intensifies the social inequalities resulting from employment.

In consequence, this chapter has argued that swimming with the tide and harnessing informal work is the only viable option. Here, we have outlined that there are currently four different discourses on what forms of informal work should be harnessed and for what purpose. First, there are some, harking back to old-style social democracy, whose approach towards poverty alleviation is grounded in a return towards full-employment. For them, the value of informal work is solely as a springboard into employment. The role of informal work is thus relevant solely in terms of its function of providing alternative routes into employment. Second, there are those who view informal work as complementary to private and public welfare provision in creating a mixed economy of welfare and third, there are those who extend this notion of informal work as complementary to the economic realm where it is seen as complementing employment. Fourth and finally, there are those who view informal work as an alternative to employment.

This review of these contrasting discourses sets the scene for the rest of the book. As shown above, the present stance is that the informal sphere is viewed as a third prong in the welfare delivery equation. When it comes to the informal sphere as a form of work, however, the perception is that such work is something to be cultivated in order to pursue formalisation so as to achieve the goal of full-employment. In this chapter, however, we have shown that there are other approaches that view the informal sphere as either a complement or an alternative to formal provision in the realms of both welfare delivery and economic production. In the rest of this book, our intention is to provide an argument for an extension to how the informal sphere is currently viewed so as to create an alternative third way to that of New Labour's employment-centred social integrationism.

To show why this is necessary, Part II provides a baseline assessment of the current role of the informal sphere in household coping practices. Part III then investigates how the informal sphere can be harnessed, especially in deprived populations that are excluded not only from employment but also from informal work relative to affluent populations, and in a way that does not lead to the creation of a dual society.

Part II

Examining poverty

Household coping capabilities and practices

4 Coping capabilities

Like all other poverty analyses (see Alcock 1997), the way in which poverty is here defined and measured is heavily embedded in a particular conceptualisation of what should be done about it. As highlighted in the introduction, we do not view poverty primarily to result from having no income (the view of the Old Left) or no morals (the New Right), nor do we view it to be chiefly the outcome of having no employment (New Labour). Instead, our view is that poverty stems from the dominance of an employment-centred paradigm and a resultant lack of capability to participate in work beyond employment. If poverty is to be alleviated and a more inclusive society built, therefore, this capacity needs to be developed. As such both the definition and measures used and solutions sought are somewhat different to previous discourses.

Unlike much previous literature on defining, measuring and solving poverty, this chapter does not focus upon income (see, for example, Bradbury 1989, Callan and Nolan 1991, Callan *et al.* 1993, Citro and Michael 1995, Deleeck *et al.* 1992, Hallerod 1995, Veit-Wilson 1987, Walker 1987). This is only central to Old Left discourses that define poverty primarily in terms of a lack of income, measure it in terms of income disparities and seek to alleviate it by increasing or redistributing incomes. To define and measure poverty using such 'input' indicators (e.g. the number of households with incomes below an arbitrary percentage of the average) is not only inaccurate but also leads to market-oriented readings and prescriptions.

This approach is inaccurate because it measures the availability of only one of the resources, albeit an important one, that a household can use to acquire necessary goods and services (i.e. income). It does not consider whether there are other resources at the disposal of the household or lacking (e.g. informal help). Take, for example, two households with the same income who both require childcare. One has no informal support to call upon and pays a formal carer. The other uses a grandparent on an unpaid basis. The former will have significantly less disposable income than the latter. Using solely income to measure poverty fails to recognise and value the contributions of such informal work in coping practices. It is based on a commodified view of the world, assuming that formal practices are the principal means by which needs are met and desires fulfilled.

Rather than measure inputs into households, in consequence, the argument here is that it is more important to consider outputs. One approach that can be adopted to do this is to define and measure poverty in terms of the ability of a household to acquire goods and services that they deem necessary. Here, poverty is a function of not just how much money a household has but whether it has the capability to acquire necessary goods and services. By defining and measuring poverty in terms of such 'outputs' or their capacity to get work completed, the contributions of work beyond employment thus become more central not only to the definition and measurement of poverty but also to its resolution. Given that all definitions and measurements of poverty, as Alcock (1997) so persuasively shows, are rooted in a particular conceptualisation of what should be done about it, the approach adopted here to defining and measuring it is no different. If the inherently prescriptive nature of this definition and measurement of poverty appears stark, it is simply because we have made no attempt to hide its normative and prescriptive foundations.

In adopting this 'outputs' approach towards defining and measuring poverty that seeks to measure capabilities to undertake activity, however, we are far from alone. Indeed, we have eminent company. It is precisely this 'capabilities' approach that is advocated by the Nobel Prize winning economist Amartya Sen in his concept of 'social capability' (Sen 1998) as well as by the principal architect of the third way, Anthony Giddens. As the latter scholar puts it,

> It is important for the centre-left to develop a dynamic, life-chances approach to inequality . . . Economist Amartya Sen's concept of 'social capability' provides the best way for social democrats to think about these issues. Equality and inequality don't just refer to income, or to the availability of material goods – the disadvantaged need to be able to make effective use of them. Policies concerned with promoting equality should be focused upon developing people's capacities to pursue their well-being.
>
> (Giddens 2002: 39)

What policies are required, therefore, to develop 'people's capacities to pursue their well-being'? To answer this, what is first required is a 'baseline' assessment of the capabilities of households to meet their needs and desires, the practices that they use to do this and the barriers that prevent them from enhancing their capabilities. Only once these have been identified can attention turn towards the policies required to develop people's capacities. This is the subject of Part II of this book. The rest of this chapter focuses upon this first issue of providing a baseline assessment of the capabilities of households to meet their needs and desires. Here, we review two surveys that have attempted to provide such assessments.

THE POVERTY AND SOCIAL EXCLUSION (PSE) SURVEY
OF BRITAIN

Is there widespread poverty in contemporary Britain? In other words, are households capable of acquiring necessary goods and services? In order to answer this question, Gordon *et al.* (2000) rank the percentage of the population identifying different adult items as 'necessary, which all adults should be able to afford and which they should 'not have to do without'. Based on this majoritarian approach to identifying necessities, they then examine the proportion of adults who do not have such items because they cannot afford them. The result is a unique insight into the capabilities of households to acquire necessary goods and services. It displays, on an item-by-item basis, the proportion of all consumers unable to afford various items that the majority deems as necessities (see Table 4.1).

In aggregate, some 42 per cent of the population in 1999 lacked one or more 'necessities', by which is meant items that the majority of the population (more than 50 per cent) believe all adults should be able to afford and which they should not have to do without. For some, of course, this is because they do not wish to possess them (e.g. 1 per cent of people do not possess and do not want a television). Many others, however, do not have these necessities because they cannot afford them. Indeed, nearly one in four (24 per cent) households lacked three or more 'necessities' because they could not afford them.

In comparison with earlier studies, the remarkable finding of this study is that the proportion of households unable to acquire goods and services that the majority deems to be necessities is worsening, rather than improving. In 1983, 14 per cent of households lacked three or more necessities because they could not afford them. That proportion had increased to 21 per cent in 1990 and to over 24 per cent by 1999.

For example, in Britain in 1999, roughly 17 per cent of households could not afford adequate housing conditions as perceived by the majority of the population. That is, they could not afford to keep their home adequately heated, free from damp or in a decent state of decoration. About 13 per cent could not afford two or more essential household goods, like a refrigerator, a telephone or carpets for living areas, or to repair electrical goods or furniture when they break or wear out. Almost 14 per cent were too poor to be able to undertake two or more common social activities considered necessary: visiting friends and family, attending weddings and funerals or having celebrations on special occasions. Some 11 per cent of adults went without essential clothing, such as a 'warm, waterproof coat', because of lack of money and around 7 per cent of the population were not properly fed by today's standards. They did not have enough money to afford fresh fruit and vegetables, or two meals a day, for example.

Who is unable to access goods and services that the majority view as necessities? Gordon *et al.* (2000) find that while 26 per cent of all people lack

Table 4.1 Perception of adult necessities and how many people lack them (all figures show percentage of adult population), 1999

Adult population (%)	Omnibus survey: items considered		Main stage survey: items that respondents	
	Necessary	Not necessary	Don't have/ Don't want	Don't have/ Can't afford
Beds and bedding for everyone	95	4	0.2	1
Heating to warm living areas of the home	94	5	0.4	1
Damp-free home	93	6	3	6
Visiting friends or family in hospital	92	7	8	3
Two meals a day	91	9	3	1
Medicines prescribed by doctor	90	9	5	1
Refrigerator	89	11	1	0.1
Fresh fruit and vegetables daily	86	13	7	4
Warm, waterproof coat	85	14	2	4
Replace or repair broken electrical goods	85	14	6	12
Visits to friends or family	84	15	3	2
Celebrations on special occasions such as Christmas	83	16	2	2
Money to keep home in a decent state of decoration	82	17	2	14
Visits to school, e.g. sports day	81	17	33	2
Attending weddings, funerals	80	19	3	3
Meat, fish or vegetarian equivalent every other day	79	19	4	3
Insurance of contents of dwelling	79	20	5	8
Hobby or leisure activity	78	20	12	7
Washing machine	76	22	3	1
Collect children from school	75	23	36	2
Telephone	71	28	1	1
Appropriate clothes for job interviews	69	28	13	4
Deep freezer/fridge freezer	68	30	3	2
Carpets in living rooms and bedrooms	67	31	2	3
Regular savings (of £10 per month) for rainy days or retirement	66	32	7	25
Two pairs of all-weather shoes	64	34	4	5
Friends or family round for a meal	64	34	10	6
A small amount of money to spend on self weekly not on family	59	39	3	13
Television	56	43	1	1
Roast joint/vegetarian equivalent once a week	56	41	11	3
Presents for friends/family once a year	56	42	1	3

Table 4.1 (continued)

Adult population (%)	Omnibus survey: items considered		Main stage survey: items that respondents	
	Necessary	Not necessary	Don't have/ Don't want	Don't have/ Can't afford
A holiday away from home once a year not with relatives	55	43	14	18
Replace worn-out furniture	54	43	6	12
Dictionary	53	44	6	5
An outfit for social occasions	51	46	4	4
New, not second-hand, clothes	48	49	4	5
Attending place of worship	42	55	65	1
Car	38	59	12	10
Coach/train fares to visit friends/ family quarterly	38	58	49	16
An evening out once a fortnight	37	56	22	15
Dressing gown	34	63	12	6
Having a daily newspaper	30	66	37	4
A meal in a restaurant/pub monthly	26	71	20	18
Microwave oven	23	73	16	3
Tumble dryer	20	75	33	7
Going to the pub once a fortnight	20	76	42	10
Video cassette recorder	19	78	7	2
Holidays abroad once a year	19	77	25	27
CD player	12	84	19	7
Home computer	11	85	42	15
Dishwasher	7	88	57	11
Mobile phone	7	88	48	7
Access to the Internet	6	89	54	16
Satellite television	5	90	56	7

Source: Gordon *et al.* (2000: Table 1).

two or more necessities, this is the case for 77 per cent of unemployed people, 61 per cent of the sick/disabled, 62 per cent of lone parents and 61 per cent of local authority tenants. More importantly, 70 per cent of people on income support could not afford these necessities, revealing that such support was in 1999 inadequate to acquire what the majority view as necessities.

This extensive research thus provides a comprehensive map of the capabilities of people and households to acquire goods and services deemed necessary by the majority of the population. In doing so, it displays the extensiveness of poverty in contemporary Britain. This study, nevertheless, adopts a majoritarian approach towards defining necessities. It treats as necessities only items that the majority deems to be necessities that adults should not have to do without. A second case study considered here, however, adopts a different approach to defining necessities in order to assess capabilities.

HOUSEHOLD COPING CAPABILITIES AND PRACTICES IN CONTEMPORARY ENGLAND

Between 1998 and early 2001, we conducted a survey of household coping capabilities and practices through 861 face-to-face interviews in eleven higher- and lower-income urban and rural English localities. Here, we first present some background information on the methodology employed to explore household coping capabilities and practices as well as the areas studied and following this, report the socio-spatial inequalities in the coping capabilities of households in contemporary England.

Methodology

Unlike the above PSE survey, this English localities survey of household coping capabilities (and practices) investigated whether households were managing to get 'necessary' work completed by examining 44 common domestic service activities (see Table 4.2). For each activity, households were asked whether it had been undertaken during the previous five years/year/ month/week (depending on the activity). If they asserted that it had not, then the interviewer asked them in an open-ended manner why not. These responses were classified according to whether the household did not see the task to be necessary and had thus not undertaken it or whether the household saw the task as necessary but had not undertaken it for some reason. The latter responses were classified by the reason given for not undertaking the task (e.g. cost, time). This allowed households to define themselves whether they saw a task as 'necessary' and whether they had completed it. Where a task had been undertaken, moreover, the assumption was made that this task was deemed necessary. Rather than define necessities through a majoritarian approach, therefore, this study adopts a self-definitional approach. Below, we report the results on household capabilities, by which we mean whether they are able to undertake tasks that they deem to be necessary.

Before examining the coping capabilities of households, however, we will here outline how the practices used by households were identified (which will be returned to in the next chapter) and the types of locality studied. To analyse household coping practices, whenever a task had been undertaken, a series of questions were posed. First, they were asked who had conducted the task (which was then classified into various categories including whether they were a household member, a relative living outside the household, a friend, neighbour, firm, landlord, etc). Second, they were asked whether they had been paid, given a gift or had been unpaid for conducting the task. And third, and if paid, whether they had been paid 'cash-in-hand' or not. On a broad level, this allowed all activity to be classified as one of four forms of work: self-provisioning, unpaid community work, paid informal work or formal employment. On a deeper level, it enabled the various sources of labour used for each form of work to be examined (e.g. whether paid informal

Table 4.2 Forty-four tasks investigated to ascertain household coping capabilities and practices

Home maintenance	Hairdressing
Outdoor painting	Administration
Indoor painting	
Wallpapering	*Making and repairing goods*
Plastering	Make clothes
Mend broken window	Knitting
Maintain appliances	Repair clothes
	Make furniture
Home improvement	Make garden
Double glazing	equipment
Plumbing	Make curtains
Electrical work	
House insulation	*Car maintenance*
Put in bathroom	Wash car
Build a garage	Repair the car
Build an extension	Car maintenance
Convert attic	
Put in central heating	*Gardening activities*
Carpentry	Indoor plants
	Outdoor borders
Routine housework	Outdoor vegetables
Do housework	Lawnmowing
Clean the house	
Clean windows	*Caring*
Spring-cleaning	Babysitting (day)
Do the shopping	Babysitting (night)
Wash clothes/sheets	Courses (e.g. piano
Ironing	lessons)
Cook the meals	Pet care
Wash dishes	

work had been conducted by a relative, friend, neighbour, firm or even household member). Once the source of labour had been identified for each task, the respondent was then asked in an open-ended manner why they had used that source of labour. This enabled the motivations for using particular sources of labour to be investigated.

Having completed this, an investigation was then undertaken of whether household members had conducted any of these 44 tasks for other households. Taking each task in turn, they were again asked whether they had conducted the tasks for another household within a particular time frame and if so, for whom, whether it was unpaid, or whether they received money or a gift and why they had conducted the task.

In order to gather information on informal work lying outside the 44 tasks surveyed, open-ended questions were asked about any other work that they had received on an unpaid or paid informal basis. In addition, and on the supply side, questions were also asked about any other work that household members had provided on an unpaid or paid informal basis. Finally, and in order to try to identify any work that respondents had conducted on a paid informal basis for firms (rather than other households), a series of open-ended probes were used.

Previous research using a similar technique reveals that when the results from households as customers and suppliers are compared, the same levels of unpaid and paid informal work are identified, meaning that the technique does not suffer from under- or over-reporting by respondents (e.g. Pahl 1984). Indeed, this was also found in this survey suggesting that the data is relatively reliable. Take, for example, paid informal work. If respondents were going to hide any form of work, especially on the supply side, then it would most likely be this type of informal work. However, and as revealed elsewhere (Williams and Windebank 2001a), when the results of the amount respondents report that they have received for paid informal work are compared with the amount paid by respondents as customers, the average amount paid is almost equal. The suggestion, therefore, is that there has been no tendency in this survey for suppliers to under-report the work that they conduct.

Besides identifying household coping capabilities and practices, the additional aim of this survey was to identify the barriers to participation in work beyond employment (the subject matter of Chapter 6). First, and as stated above, this was indirectly investigated by asking households why they had not undertaken a task whenever it had not been completed. Second, however, the barriers to participation in informal work were also more directly investigated. Using the main barriers to participation in informal work identified in previous surveys both in the UK and abroad (e.g. Pahl 1984, Thomas 1992), attitudinal scales were used to test whether households considered these factors were a constraint on their participation in first, self-provisioning and second, paid informal exchange. Using a range of statements such as 'I would do more "cash-in-hand" work if I did not fear being caught', respondents were asked whether they strongly agreed, agreed, neither agreed nor disagreed, disagreed or strongly disagreed with such statements. A third and final way in which these barriers to participation in informal work were investigated was through open-ended questions about what prevents them from doing first, more self-provisioning, second, more unpaid work for others outside the household and third, more 'cash-in-hand' work. Chapter 6 will report the results.

To enable household coping capabilities and practices as well as the barriers to participation to be considered according to household type, the final stage of the interview sought background information. This included the number of adults and children in the household, how long they had lived in the area, the job histories of the household members, the current employment status of each member, the gross household income, household tenure and the ethnic composition of the household. Two additional questions were also asked. The first was about the extent of the household's kinship networks in the locality with regard to whether any household member had any grandparents, parents, cousins, uncles and aunts, brothers and sisters or children living in the city (besides those living in the household with them). The second was about the nature of their social networks (asked only in the rural surveys). This requested information about whether they perceived themselves as belonging to any formal or informal network, association or group and, if so, whether they had been able to draw upon members of this group to help them in the recent past. If they responded positively, then they were asked about the nature of the help provided. Finally, and as is the case with all interviews, much additional information was provided in each interview that was pertinent to the aims of the survey. Immediately after each interview, therefore, the researcher wrote up the additional information collected in the form of a cameo of the household as well as specific information that might be useful.

Here in this chapter, we will report the household coping capabilities, followed by household coping practices in Chapter 5 and barriers in Chapter 6. Before doing so, however, it is necessary briefly to introduce the areas studied.

Localities studied

This survey method was applied in 11 contrasting English localities (see Table 4.3). On the one hand, the decision was taken to study higher- and lower-income populations. First, this was due to the common perception that household coping capabilities are likely to vary significantly according to income. Second, it was due to the vast body of work reviewed in Chapter 2 that displays how the use of work beyond employment is strongly tied to level of affluence. On the other hand, urban and rural areas were surveyed. This was because of the growing recognition that poverty is not solely an urban phenomenon. It is also widespread in rural areas (e.g. Countryside Agency 2001, Shucksmith 2000) where different problems confront households so far as capabilities as well as practices and barriers to participation in informal work are concerned.

To ensure that the data collected in each locality was representative of the area, in each and every locality the same sampling procedure was used. The researcher called at every nth dwelling in each street, taking into account multiple units in some dwellings. Depending on the size of the locality and the number of interviews sought, this obviously varied to some extent in each area. If there was no response, then the researcher called back once. If there was still no response and/or they were refused an interview, then the $(n + 1\text{th})$ house was surveyed (again with one call back) followed by the $(n - 1\text{th})$ dwelling, $(n + 2\text{th})$ and so forth. This provided a representative sample of the locality and prevented any skewness in the sample towards certain tenures, types of dwelling and different parts of the locality being interviewed rather than a representative sample of the whole area.

Table 4.3 Areas studied

Area-type	Locality	Number of interviews	Description of area
Affluent rural	Fulbourn, Cambridgeshire	70	Affluent 'picture postcard' rural village in high-tech sub-region
Affluent rural	Chalford, Gloucestershire	70	Affluent rural village in Cotswolds
Lower-income rural	Grimethorpe, South Yorkshire	70	Ex-pit village with very high unemployment
Lower-income rural	Wigston, Cumbria	70	Deprived rural village with one factory dominating the local labour market
Lower-income rural	St Blazey, Cornwall	70	Deprived rural locality in a tourist region
Affluent suburb	Fulwood, Sheffield	70	Affluent suburb in south-west Sheffield
Affluent suburb	Basset/Chilworth, Southampton	61	Only affluent suburb within the city of Southampton
Lower-income urban	Manor, Sheffield	200	'Sink' social housing project with high unemployment
Lower-income urban	Pitsmoor, Sheffield	200	Deprived inner city area in deindustrialising city with high levels of private sector rented accommodation and high unemployment
Lower-income urban	St Mary's, Southampton	200	Deprived inner city locality in affluent southern city with high levels of private sector rented accommodation and high unemployment
Lower-income urban	Hightown, Southampton	200	Deprived social housing project with high unemployment

It was important, moreover, for a representative sample to be achieved, that interviewing did not take place solely during daytime hours on weekdays. In order for multiple-earner households to be captured, interviews also occurred during the early evening and weekends. To prevent 'cold calling', all of the target households had a covering letter put through their door a day or so prior to the researcher calling in order for them to receive some information about the nature of the interview and to hopefully increase the response rate. This letter described the research as being interested in finding out how households manage to get everyday tasks completed and what prevents them being able to do more for themselves and others.

Below, we report the results on household coping capabilities, starting with the socio-economic inequalities and then the spatial disparities.

Socio-economic disparities in household coping capabilities

To analyse how the capabilities of households to undertake tasks that they view to be necessary vary socio-economically, the results from the whole sample of 865 households are here reported. This reveals, as might be expected, significant variations in the coping capabilities of households according to their income level and attachment to the formal labour market (see Table 4.4).

Of the 44 domestic tasks investigated, no-earner and lower-income (i.e. with a gross household income of less than £250 per week) households complete a narrower range of tasks than multiple-earner and higher-income (i.e. gross households incomes of £250 or more per week) households. On its own, this tells us little about the capability of households to perform necessary work. After all, households might not have wished to undertake many of these tasks.

The next column of Table 4.4 thus takes into account whether households wished to do tasks and examines what proportion of all tasks that they wished to undertake were not completed. This reveals that jobless and lower-income households not only conduct a smaller range of tasks but also are less able than their employed and higher-income counterparts to get tasks that they deem necessary completed. Indeed, jobless households had been unable to complete 43.1 per cent of the tasks that they perceived as necessary, compared with 28.7 per cent in multiple-earner households. This lesser ability of jobless households to get necessary work completed, nevertheless, might be due to a whole range of hypothetical factors ranging from a lack of time to an active social life that they prioritise over getting tasks completed around the home to maintain their material standard of living.

For this reason, why households had been unable to get these tasks completed was investigated. This reveals that amongst jobless households,

Table 4.4 Household coping capabilities: socio-economic disparities

	% of 44 tasks completed	*Of tasks need to do, % not done*	*% of these not done because cannot afford*
All households	50.5	34.1	82.3
No. of earners			
Multiple	60.1	28.7	60.0
Single	54.3	32.7	71.1
None	44.2	43.1	90.0
Gross household income			
> £250 per week	56.1	29.6	66.4
< £250 per week	45.3	40.1	89.6

in 90 per cent of cases where a necessary task had not been done, this was because they could not afford to undertake it. In multiple-earner households, in contrast, where a wider range of tasks are completed and households manage to complete a greater proportion of the tasks that they deem necessary, just 60 per cent of the outstanding necessary tasks were not conducted due to affordability.

Such an analysis, however, congregates together very different household types. Take, for example, multiple-earner households. A very important distinction needs to be made between higher-income multiple-earner households and what we here call multiple-earner 'working poor' households (i.e. multiple-earner households in which the gross household income is less than £250 per week). When this is done and their contrasting coping capabilities are analysed, it is revealed that inserting people into employment is not always a solution to poverty, measured in terms of household coping capabilities.

Examining the multiple-earner 'working poor' households shows that entry into employment does not enhance their coping capabilities relative to their jobless working-age counterparts. These 'working poor' households fail to undertake 48.9 per cent of the tasks that they deem as necessary, compared with 43.3 per cent in jobless working-age households. As such, the apparently wealthy multiple-earner household category contains a group of 'working poor' who are worse off than their jobless counterparts in terms of their coping capabilities. The important lesson, therefore, is that entry into employment does not resolve poverty, measured in terms of coping capabilities, when the formal jobs taken are low paid. This will be returned to in Chapter 6.

The differentials in these socio-economic disparities, moreover, are not universal in their magnitude. They significantly vary spatially. Not only does 'work (employment) pay' more in some areas than others, but so too is it better to be unemployed in some places than others.

Inter-urban disparities in household coping capabilities

To what extent does the capacity of households to perform necessary work (i.e. our definition of poverty) vary between the relatively affluent southern city of Southampton and the relatively deprived northern city of Sheffield? And why is this the case? It might be assumed that households in the relatively affluent city of Southampton are more able to get necessary work completed than their relatively deprived northern counterparts. Such an assertion about the spatial variations in household coping capabilities assumes that because Southampton has a higher employment rate, higher wage levels and household income is greater, so too will be their ability to get necessary work completed. Here, however, we reveal that the picture is far more complex and by no means as clear-cut as such a portrayal assumes.

As Table 4.5 displays, of the 44 everyday tasks investigated, Southampton households managed to complete a narrower range than Sheffield households.

Table 4.5 Household coping capabilities: comparison of Southampton and Sheffield

	% of 44 tasks completed	Of tasks need to do, % not done	% of these not done because cannot afford
All urban areas	48.8	36.8	84.6
Southampton	47.7	38.7	86.4
Sheffield	50.0	34.8	82.9

As stated above, this tells us little about the capability of households to perform necessary work since households might not have wished to undertake many of these tasks. The next column of Table 4.5 thus investigates the proportion of the tasks that they wished to undertake that were completed. This reveals that Southampton households not only conduct a smaller range of tasks but also are less able than their Sheffield counterparts to get tasks that they deem necessary completed. Indeed, Southampton households had been unable to complete 38.7 per cent of the tasks that they perceived as necessary, compared with 34.8 per cent in Sheffield. Examining why this is the case reveals that amongst Southampton households, in 86.4 per cent of cases where a necessary task had not been done, this was because they could not afford to undertake it. In Sheffield, in contrast, where a wider range of tasks are completed and households manage to complete a greater proportion of the tasks that they deem as necessary, a slightly lower share of the outstanding necessary tasks were not conducted due to affordability.

Consequently, households in Sheffield manage to complete a wider range of tasks, get more of the necessary tasks undertaken and affordability is less often the reason for not doing outstanding tasks than in Southampton. It appears, therefore, that the coping capabilities of households in Southampton are weaker than in Sheffield. Is it the case, however, that the coping capabilities of all households are weaker in Southampton than in Sheffield?

Inter-urban variations in the coping capabilities of jobless households

Given that jobless households claiming benefits receive the same amount of money wherever they live, it might be assumed that government bestows on all such households an equal opportunity to 'enjoy' the same standard of living. This applies to both working-age unemployed people as well as those who have spent a life working and are now dependent on state retirement benefits. Examining Table 4.6, however, reveals that the capabilities of these households differ significantly between the two cities. In Sheffield, jobless households manage to complete not only a wider range of the 44 tasks than Southampton jobless households (47.8 per cent compared with 42.4 per cent) but also a greater proportion of the tasks deemed necessary (58.9 per cent

Table 4.6 Coping capabilities of jobless households: inter-urban disparities

	% of 44 tasks completed	Of tasks need to do, % not done	% of these not done because cannot afford
All urban areas	45.1	42.8	89.0
Southampton	42.4	44.4	98.0
Sheffield	47.8	41.1	83.0

compared with 55.6 per cent). Affordability, moreover, is less often the reason for not doing the outstanding tasks in Sheffield than in Southampton.

To explain why jobless households in Southampton appear to be worse off than their Sheffield counterparts, it is perhaps necessary to consider whether this is due to the differential cost of living in the two cities. Despite the fact that households reliant on social protection payments (e.g. pension payments, disability benefits, unemployment benefits) receive the same wherever they live, crude data exists to suggest that there is a 9.5 per cent difference in the cost of living between Southampton and Sheffield (Reward Group 1999). This may well explain why Southampton jobless households are worse off than their Sheffield counterparts in terms of their ability to perform necessary tasks. However, it is not only that there is a gap in the coping capabilities of jobless households between Southampton and Sheffield. There is also a gap in the extent to which 'work (employment) pays' between the two cities.

Inter-urban variations in the gap between the coping capacities of employed and jobless households

Table 4.7 reveals that the benefit of having somebody in employment in a household rather than everybody jobless is greater in Southampton than Sheffield. Put another way, 'work (employment) pays' to a greater extent in Southampton than in Sheffield. Comparing the ability of multiple-earner and no-earner households to get work completed, for example, Southampton multiple-earner households completed 49 per cent more tasks than no-earner households but in Sheffield this difference was just 14 per cent. In Southampton, moreover, there was a difference of 16.1 percentage points between multiple- and no-earner households in the proportion of tasks deemed necessary that were completed, whilst in Sheffield, the difference between multiple and no-earner households amounted to just 10 percentage points. The difference in the coping capabilities of employed and jobless households is thus wider in Southampton than in Sheffield.

Why is it the case, therefore, that 'employment pays' to a greater extent in Southampton than in Sheffield, measured in terms of household coping capabilities? This might be tentatively explained in terms of the inter-

Table 4.7 Variations in coping capabilities of employed and jobless households: by city

	% of 44 tasks completed	Of tasks need to do, % not done	% of these not done because cannot afford
Southampton			
Multiple-earner	63.1	28.3	60.3
Single-earner	58.7	33.2	70.2
No-earner	42.4	44.4	98.0
Sheffield			
Multiple-earner	54.5	30.1	60.0
Single-earner	50.7	31.6	72.2
No-earner	47.8	41.1	83.0

relationships between wage rates, cost of living and social payments. On the one hand, Southampton jobless households receiving nationally determined social payments (e.g. pensions, welfare benefits) are worse off than their Sheffield counterparts due to the higher cost of living in this southern city. On the other hand, employed households are better off in Southampton than in Sheffield, measured in terms of the ability to get necessary work completed, because the difference in the wage rates between Southampton and Sheffield exceed the difference in the cost of living between the two cities. We have already stated that the difference in the cost of living appears to be around 9.5–10 per cent. Gross average weekly wage rates, however, are 13.5 per cent higher in Southampton than in Sheffield. They are £384.00 in Southampton compared with £338.20 in South Yorkshire (New Earnings Survey 1998). The result is a wider 'wealth gap' in Southampton than in Sheffield between those in employment and those dependent on state social payments.

Such an explanation, however, has to remain tentative. Data is first required on *real* wage rates and *real* household incomes before it can be stated with any certainty and for this to occur, accurate measures of the variations in the cost of living are required. The data reported here, nevertheless, certainly supports the notion that there are spatial variations in the extent to which 'employment pays'. That is, the gap between employed and unemployed households in terms of their ability to get necessary work completed (as defined by respondents) is far greater in Southampton than in Sheffield.[1]

In sum, examining the inter-urban disparities in household coping capabilities, it has here been revealed that first, one is better off living in a jobless household in Sheffield than in Southampton and second, 'employment pays' in Southampton more than in Sheffield. The former may be because despite variations in the cost of living, there are flat-rate social protection payments (e.g. for the unemployed, pensioners), whilst the latter may be because the difference in the wage rates between Southampton and Sheffield exceed the difference in the cost of living between the two cities.

Intra-urban disparities in household coping capabilities

Spatial disparities in coping capabilities, however, do not only occur at the inter-urban level. Significant variations also prevail at the intra-urban level. To show this, we here investigate the differences between lower-income urban neighbourhoods, namely council estates and deprived inner city neighbourhoods, and affluent suburbs.

Although it is widely assumed that the coping capabilities of households in affluent suburbs will be greater than in lower-income urban neighbourhoods, there has until now been little attempt to explore the magnitude of this differential. Table 4.8 provides some indication. It reveals the extent to which, in both cities, households in affluent suburbs are more able to perform necessary work than those in lower-income urban neighbourhoods. This 'wealth gap' between higher- and lower-income urban neighbourhoods in terms of the ability of households to perform necessary work should be of no surprise.

More interesting is that this 'wealth gap' between higher- and lower-income neighbourhoods in terms of coping capabilities is greater in Southampton than in Sheffield (see Table 4.8). In Southampton, households in higher- and lower-income neighbourhoods were unable to undertake 15.2 per cent and 44.3 per cent of necessary tasks respectively (a gap of 29.1 percentage points). In Sheffield, meanwhile, this difference was much narrower with higher- and lower-income neighbourhoods failing to complete 21.9 per cent and 38.4 per cent of tasks respectively (a gap of just 16.5 percentage points).

Again, this can tentatively be explained by examining the interrelationship between wage levels, cost of living and social payment levels. In Southampton, lower-income neighbourhoods, where those depending on state benefits are concentrated, suffer relative to their Sheffield counterparts

Table 4.8 Household coping capabilities: intra-urban disparities

	% of 44 tasks completed	Of tasks need to do, % not done	% of these not done because cannot afford
All urban areas	48.8	36.8	84.6
Southampton	47.7	38.6	86.4
Council estate	47.4	39.5	86.0
Inner city	43.1	49.1	97.0
Both lower-income areas	45.3	44.3	92.0
Affluent suburb	57.3	15.2	64.0
Sheffield	50.0	34.8	82.9
Council estate	48.3	35.9	78.0
Inner city	49.7	40.5	92.0
Both lower income areas	49.0	38.4	85.0
Affluent suburb	53.3	21.9	76.0

(displayed in the lower proportion of necessary work completed) due to the higher cost of living. In affluent suburbs where more are employed, meanwhile, households benefit in Southampton from wage levels that outstrip the higher cost of living, resulting in a greater ability to get necessary work completed (as displayed in Table 4.8). The outcome is this greater 'wealth gap' between higher- and lower-income areas in Southampton than in Sheffield.

It is not only at the inter- and intra-urban level, however, that spatial variations can be identified in coping capabilities. There are also significant urban–rural variations.

Urban–rural variations in household coping capabilities

As Table 4.9 displays, despite the growing body of work on the prevalence of rural poverty (see, for example, Shucksmith 2000), the coping capabilities of rural households appear to be greater than their urban counterparts. This applies whether one examines urban and rural areas as a whole or deprived rural and urban areas. In all these area-types, rural households are more able to get a greater proportion of the work that they view as necessary completed than their urban counterparts.

At first glance, this finding about the greater coping capabilities of rural households seems to run counter to the explanations being put forward in this chapter. There is now much evidence that wages are lower in rural areas (Chapman *et al.* 1998, Shucksmith 2000) and the cost of living higher (Cabinet Office 2000). Following the logic so far imparted in this chapter, then households living in rural areas should be less capable of undertaking necessary work than their urban counterparts. However, the evidence above suggests that this is not the case.

Our explanation is that lower wage levels in rural areas coupled with higher living costs do not reduce coping capabilities because informal work in rural areas is higher. Such work is used to compensate for the inability of households to participate in the formal realm. Put another way, although having a lower ability to draw upon formal sources in their coping practices, rural

Table 4.9 Urban–rural variations in household coping capabilities

	% of 44 tasks completed	Of tasks need to do, % not done	% of these not done because cannot afford
All urban areas	48.8	36.8	84.6
All rural areas	53.6	33.3	81.1
Lower-income urban	47.1	41.4	89.0
Lower-income rural	53.4	38.8	84.0
Affluent urban	55.2	18.7	70.0
Affluent rural	53.9	23.0	62.0

Table 4.10 Urban–rural variations in coping capabilities of jobless households

	% of 44 tasks completed	Of tasks need to do, % not done	% of these not done because cannot afford
All urban areas	45.1	42.8	89.0
All rural areas	47.6	46.8	93.3
Lower-income urban	44.1	50.4	93.0
Lower-income rural	45.2	50.8	97.2
Affluent urban	50.7	23.8	84.8
Affluent rural	49.2	29.8	71.2

households are able to more than compensate by their ability to draw upon informal work. Evidence that this is the case will be provided in the next chapter. For the moment, we leave it to the side.

Before concluding our review of the urban–rural variations, it is important to note that although rural populations as a whole have greater coping capabilities than their urban counterparts, this is not the case for at least one particular social group: jobless households. Examining Table 4.10, we see that the coping capabilities of jobless households in rural areas are lower than those of their urban counterparts. Not only are rural jobless households less able than their urban counterparts to get necessary work completed but financial constraints are more prominent in their reasons for not being able to do this work than is the case in jobless urban households.

In part, this is an outcome of the fact that rural jobless households suffer a higher cost of living, notably in the realms of food and transport (Cabinet Office 2000), even though they receive the same flat-rate social benefit payments as their urban counterparts. The result is that they are unable to make their money go so far as their urban equivalents. Unlike those rural households with people in employment, however, these jobless households are unable to draw upon informal practices to the same extent in order to mitigate their circumstances (see next chapter). The result is that the higher cost of living results in them suffering lower coping capabilities than their urban counterparts.

CONCLUSIONS

In Part I of this book, we highlighted some rationales for advocating an alternative third way approach that seeks to alleviate poverty and build a more inclusive society by developing the capabilities of people to participate in work beyond employment. In this opening chapter of Part II, a 'baseline' assessment has been provided of the current capabilities of households to acquire necessities. This has revealed not only how a significant proportion of households are unable to acquire items that either others or they deem to be necessities, but also the socio-spatial disparities in coping capabilities.

Having identified these variations in capabilities, the next chapter turns its attention to the practices' households currently use in order to acquire necessities. In so doing, it will be shown that if poverty is to be alleviated and a more inclusive society constructed, it will be important to harness the ability of households to participate in informal economic activity.

Note

1   For a fuller review of the inter-urban disparities in the coping capabilities of jobless households and the uneven extent to which 'work pays', as well as the policy implications, see Williams (2001a).

5 Coping practices

If coping capabilities significantly vary between deprived and affluent populations, is it nevertheless the case that lower-income populations manage to compensate for their lack of income by engaging in informal modes of production? To answer this, the particular mix of practices that households use to get necessary work completed, or what are sometimes referred to as their 'household work practices' (e.g. Leonard 1994, 1998, Meert *et al.* 1997, Nelson and Smith 1999, Pahl 1984, Warde 1990, Williams 2002a, 2002b), are here reported.

Examining the results of the 861 household interviews conducted in higher- and lower-income English urban and rural areas, this chapter will reveal that households heavily rely on informal practices in order to get necessary work completed in the domestic realm. However, little evidence is found to support the assertion of the 'marginality thesis' that informal work is more widespread amongst lower-income populations (see Chapter 2). Instead, informal modes of production are found to reinforce, rather than reduce, the socio-spatial inequalities produced by the formal labour market. Those excluded from the formal labour market are also the least able to draw upon informal practices in order to enhance their coping capabilities.

This finding that affluent populations conduct more informal work leads to questions about whether such work is conducted purely out of necessity or whether other rationales are involved. To evaluate this, the motives for different populations engaging in self-provisioning, unpaid community exchange and paid informal work are analysed in turn. This will reveal that many lower-income households would like to engage in informal modes of production in order to enhance their capacities but are currently unable to do so. The next chapter will then explore why this is the case. To begin, however, it is necessary to highlight the extent to which households draw upon informal work in their coping practices.

Table 5.1 Household coping practices: by type of area

% of tasks last conducted using:	Self-provision-ing	Unpaid community work	Paid informal exchange	Formal employ-ment	χ^2
All areas	71.4	5.0	5.5	18.1	
Lower-income rural areas	67.1	7.2	5.6	19.9	89.76
Higher-income rural areas	63.4	7.5	4.1	24.0	28.88
All rural areas	65.6	7.3	5.0	21.6	56.67
Lower-income areas – Southampton	74.8	3.6	4.4	17.3	38.98
Higher-income suburb – Southampton	71.3	1.9	6.5	20.3	29.88
Lower-income areas – Sheffield	77.4	3.9	5.4	13.3	174.19
Higher-income suburb – Sheffield	72.8	1.9	11.2	14.1	29.86
All urban areas	75.2	3.4	5.8	15.6	54.37

$\chi^2 > 12.838$ in all cases, leading us to reject H_0 within a 99.5 per cent confidence interval that there are no spatial variations in the sources of labour used to complete the 44 household services.

THE PREVALENCE OF INFORMAL WORK IN HOUSEHOLD COPING PRACTICES

To what extent has the formal sphere penetrated the domestic realm? Examining the sources of labour used in contemporary England by the 861 households interviewed, the finding is that 18.1 per cent, or less than one in five, of the 44 domestic tasks were last completed using formally employed labour. This suggests that the domestic services sector is far from totally commodified. Indeed, in a fully formalised economy, the household services sector would be five times larger than its present size. However, this would require a formalisation of the 71.4 per cent of tasks last conducted on a self-servicing basis, the 5.0 per cent of tasks last conducted using unpaid labour from outside the household and the 5.5 per cent of tasks using paid informal labour (see Table 5.1). Whether this is possible will be returned to below. Here, it is simply important to note that the formalisation of the domestic realm is far from the reality. Informal economic activities remain the dominant source of labour used in household coping practices.

Household work practices, however, are not uniform across space. As Table 5.1 reveals, households in higher-income localities have both more formalised and more monetised coping practices than those in lower-income localities.

They use formal or informal paid labour to undertake a greater proportion of their domestic services. From this finding, it might be assumed that as household income rises, a greater proportion of the domestic workload is externalised to formal labour. Lower-income areas, in contrast, tend not only to rely to a greater extent on self-provisioning in their coping practices but also to display a higher proclivity to draw upon unpaid community work. Superficially, these findings are not controversial.

More controversial is our finding concerning the reasons for these spatial variations in coping practices. For some, it might be assumed that these spaces of informality are the product of a new post-Fordist regime of accumulation that is offloading social reproduction functions from the formal sphere back on to the informal sphere (e.g. Castells and Portes 1989, Lee 1999, Portes 1994). The breakdown of the post-war economic regulations and welfare state through de-regulation and flexibilisation of social relations of production, and the transferring of social services to private and communal hands (Gershuny and Miles 1983, Pahl 1984) could be seen to have led to a process of informalisation. Informal economic activities have expanded to occupy spaces of production (and reproduction) previously covered by market relations and state subsidies. In this view, it is thus the contradictions inherent in the formalisation process itself that have led to the informalisation of spheres of social reproduction. In order to compete in the global economy, advanced economies have had to reduce social costs (see European Commission 2000a). Activities associated with social reproduction have been decanted from the formal sphere into the informal sphere, resulting in a shifting balance between the formal and informal realms. The result is a socio-spatial polarisation of coping practices. The social reproductive functions of those excluded from the formal sphere are being offloaded on to the informal sphere.

Although this structural explanation is one way of understanding these socio-spatial differences in the degree of formalisation, our argument is that economic essentialism is alone insufficient to explain the uneven contours of informalisation. Such an argument ignores that although affluent populations are more formalised and monetised in their coping practices than deprived populations, they also undertake a disproportionate amount of the total informal work that is conducted in society. Examining the total sample of 861 households, this study finds that contrary to the marginality thesis, multiple-earner households, despite constituting only 24 per cent of the households surveyed, conduct 32 per cent of all tasks carried out on an informal basis. No-earner households, meanwhile, comprising 51 per cent of the households surveyed, conduct just 41 per cent of all tasks undertaken informally.

Breaking this down into the various types of informal work, multiple-earner households, although just 24 per cent of all households surveyed, undertake 35 per cent of all self-provisioning, 22 per cent of all unpaid community work and 37 per cent of all paid informal exchange. No-earner households, meanwhile, although representing 51 per cent of all households surveyed, conduct just 40 per cent of all self-provisioning, 54 per cent of unpaid exchange and 28 per cent of paid informal exchange.

The above economic explanation, in consequence, does not suffice. It does not explain why higher-income households with more formalised and monetised coping practices also conduct a greater proportion of all informal work. If this is to be explained, the argument put forward in the rest of this chapter is that economic constraints are not the sole reason for using informal practices and, as such, are insufficient as an explanation for the prevalence and distribution of informal work. Indeed, in only 43 per cent of cases where formal labour was not used, did households give cost or economic necessity as their rationale. Other rationales beyond economic necessity are thus required in order to explain the socio-spatial disparities in informal work. As we shall now show, it is the contrasting ways in which structure and agency influence its advance in different populations that need to be grasped.

To provide a more in-depth understanding of the reasons for the persistence of informal practices, we thus here examine the socio-spatial variations in the reasons for engaging in first, self-provisioning, second, unpaid community work and third and finally, paid informal exchange.

PARTICIPATION IN SELF-PROVISIONING

Is it the case that as household income rises, a greater proportion of the household workload is undertaken by formal labour? In other words, are higher-income populations more likely to employ gardeners, nannies, cleaners and formal craftspeople to undertake work? And does this externalisation result in a net reduction in self-provisioning for these households?

The finding of this study is that although a greater proportion of the domestic workload is formalised in higher-income households, they still engage in a wider array of self-provisioning than lower-income populations. Multiple-earner households constitute just 24 per cent of the households interviewed but they conduct 28 per cent of all self-provisioning tasks. No-earner households (51 per cent of the sample) conduct 49 per cent of self-servicing activity. Consequently, although there is a relative shift of the workload in higher-income households towards the formal sphere, this does not mean that the absolute amount of self-provisioning decreases in such households. Rather, the contracting out of some routine self-provisioning enables these households to engage in more non-routine self-provisioning on a freely chosen basis, as will be more fully explicated below.

Lower-income households, meanwhile, engage in a narrower range of self-provisioning than their higher-income counterparts. To explain this, it is necessary to explore both the composition of these households and the type of work that they undertake. On the issue of their composition, lower-income households are more likely to be composed of a sick/disabled person or the retired, to be in rented accommodation and to be jobless than higher-income households. As such, they are less likely to possess the combination of money, tools, knowledge, practical skills and physical ability, and/or to have the responsibility to do the work. Lower-income households thus conduct largely routine tasks (e.g. everyday housework) or essential tasks when absolutely necessary (e.g. when the electricity fails, a water leakage occurs, the cooker breaks down).

For higher-income households, meanwhile, a greater proportion of tasks is voluntarily chosen (e.g. wallpapering, painting and various other do-it-yourself activities) and conducted out of choice rather than necessity. Given that more voluntary tasks often require less skill than more necessary tasks (e.g. when a cooker breaks down or a roof leaks), lower-income households are relatively heavily reliant on buying-in labour (either formally or informally) to get a greater proportion of their total workload completed than higher-income households. This is simply due to the type of jobs that they conduct (see also Davidson *et al.* 1997, Leather *et al.* 1998).

In contemporary English society, the outcome is that self-provisioning is not a substitute for money income. It is the companion of money income. This produces a new kind of poverty. Those without money are also unable to engage in self-provisioning, reinforcing rather than reducing their situation.

Why, therefore, do households engage in self-provisioning? Are their motives the same amongst all social groups and in all places? Or do they vary socio-spatially? To answer this, each time a task was undertaken on a self-servicing basis, interviewees were asked why the task had been conducted using this form of labour. The finding, as Table 5.2 reveals, is that not all this activity was undertaken out of economic necessity. Indeed, just 44 per cent of self-provisioning in the lower-income neighbourhoods was primarily motivated by economic necessity and merely 10 per cent in the higher-income neighbourhoods. Put another way, well over half (56 per cent) of all self-servicing activity was conducted for non-economic reasons and/or out of preference in lower-income areas and 90 per cent of such activity in higher-income areas.

First, that is, some 18 per cent of all self-servicing in lower-income localities and 37 per cent in higher-income areas was undertaken because it was felt to be *easier* to do the task on a self-servicing basis than to externalise the work. This applied to a wide range of activity ranging from routine housework to home improvement and maintenance. When it was stated that routine housework tasks were undertaken on a self-servicing basis because it was easier, this was explained in terms of the availability of domestic technologies (e.g. washing machines, dishwashers) and/or the problems in finding 'trustworthy' people to whom the tasks could be externalised. Indeed, it was often stated that there had been the demise in the prevalence of informal social networks through which cleaners and so forth could be found. The formal domestic service firms that had arisen to replace them, meanwhile,

Table 5.2 Reasons for conducting domestic tasks on a
self-servicing basis

Reason for using self-provisioning (per cent)	Lower-income areas	Higher-income areas
Economic	47	7
Ease	18	37
Choice	21	24
Pleasure	14	32

were seen as expensive and the labour used of poor quality and not subject to detailed police checks for their trustworthiness. In short, interviewees did not want to use them or let them into their home. So far as home improvement and maintenance tasks were concerned, similar rationales prevailed. It was often stated for example, that tradespeople were difficult to contact and even more difficult to persuade to come to do a job so it was often easier simply to do the job oneself than rely on them. In other words, a widespread distrust of professional tradespeople led households to do the work themselves.

Second, however, in some 21 per cent of cases where self-servicing was last used in lower-income neighbourhoods and 24 per cent of cases where it was last used in higher-income areas, self-servicing was the first *choice* of households rather than a last resort resulting from financial necessity. The perception was that by doing a task themselves, they could create an end product that was of a higher quality than would be otherwise the case or that the end product could be individualised in a way that suited their needs and/or desires. It displays that 'a largely unintended effect of a highly individualised and marketised society has been the intensification of social practices which systematically "evade the edicts of exchange value and the logic of the market"' (Urry 2000: 146). Reflected in popular do-it-yourself home decorating television programmes such as *Changing Rooms* that both cater to and fuel the desire for such self-servicing, this pursuit of individuation has led to the growth of self-servicing as a 'chosen space'.

Third and finally, self-servicing was often used due to the *pleasure* people got from doing the work themselves. Indeed, this reason accounted for some 14 per cent of cases where self-servicing was last used in lower-income neighbourhoods and 32 per cent of cases where it was last used in higher-income areas. This rationale, moreover, did not purely relate to the more autonomously chosen do-it-yourself projects such as home improvement, although it was relatively concentrated in this realm. A significant minority did routine housework themselves because of the pleasure they received from this activity. More often than not, these were employed women whose rationale entailed comparing the outputs of their housework with the products of their employment. As one employed woman put it,

> At least you are doing something productive and you can see what you have achieved at the end of it, unlike my job.

Or as another employed woman put it,

> Compared with my job, housework's great. You get a real buzz when you look at it when you have finished and see what you have achieved.

Although some non-employed women shared this view of routine housework as pleasurable, these were mostly older women, often of retirement age. For most respondents, however, it was non-routine self-provisioning that was perceived as pleasurable and this is the reason why pleasure was more heavily cited as a reason for self-servicing in the higher-income areas where a greater proportion of such non-exchanged work is of the non-routine variety.

In consequence, this study displays that the persistence of self-provisioning in the realm of domestic service provision is not simply the result of the financial inability of households to formalise these tasks. Households often prefer to undertake activities themselves due to ease, choice or pleasure. Indeed, this study from the UK is not alone in arriving at this finding. In Denmark, Sundbo (1997) finds that just 36 per cent of households would consider externalising spring-cleaning, 33 per cent cleaning, 27 per cent gardening, 21 per cent advice about economic matters, 20 per cent household repairs, 17 per cent care of sick children, 13 per cent shopping, 13 per cent care of children in general and 12 per cent the provision of meals. Two-thirds prefer to do these jobs themselves under all circumstances, never mind what price is charged. As such, a strong 'culture of resistance' is identified to the externalisation of activity both in this study and in the only other known investigation so far conducted.

The degree to which 'cultures of resistance' prevail over economic necessity in determining the extent of self-provisioning, nevertheless, differs across populations. Put another way, populations have contrasting preference/necessity ratios. In higher-income populations, a larger share is conducted out of choice than necessity. Indeed, despite being financially able to formalise activity, they choose only to formalise a narrow range of mostly routine, mundane and repetitive tasks and use some of the time released to engage in more non-routine, creative and rewarding self-servicing activity that they conduct out of preference rather than necessity. In lower-income populations, however, where self-servicing is composed mostly of essential and/or routine tasks, necessity tends to prevail to a greater extent.

It appears, therefore, that self-provisioning helps secure a definition of self that would be undermined by an exclusive reliance on paid employment. People guard against externalising activity in order to forge forms of self-identity and worth through this type of activity. As such, self-provisioning provides one nexus through which people forge and display who and what they are and wish to be. The outcome is that the productive functions of the household, far from disappearing under capitalism, seem to be retained as an act of resistance to the notion of defining who and what one is through employment alone (see Chapter 1).

Of course, just as in the formal labour market, the capabilities of people to use self-provisioning in this manner vary and the sizes of the rewards differ. Amongst lower-income households, self-provisioning is more routine and carries smaller material and psychological benefits. Such activity is compelled more by necessity than choice or an interest in the intrinsic satisfactions. It is much less about displaying an ability to manage on one's own and one's creative side and is much more about having to manage on one's own. In an already overburdened life where economic constraints severely impinge on one's opportunities to display who and what you are, self-provisioning represents, similar to formal employment, yet another sphere of constrained opportunities where choice is lacking and one's creative desires are stifled.

For more affluent populations, however, self-provisioning represents a sphere where needs can be fulfilled, capabilities developed and creative desires expressed, resulting in large material and psychological benefits. In consequence, just as the material and psychological rewards of formal employment are unevenly distributed, so too there are inequalities in self-provisioning. Until now, however, how this can be resolved has seldom been considered in economic and social policy. Although there is much discussion of inequalities in the sphere of employment and how the rewards can be more evenly distributed, little attention has been paid to the sphere of self-servicing. How this might be achieved will be returned to in Part III.

PARTICIPATION IN UNPAID COMMUNITY EXCHANGE

If self-provisioning reinforces, rather than reduces, the socio-economic disparities produced by the formal labour market, is this also the case so far as unpaid community exchange is concerned? In the UK, two mutually exclusive narratives exist that answer this question in different ways. On the one hand, some accounts conceptualise lower-income populations as solidaristic communities who have a greater tendency to help each other out. This harks back to studies conducted over a quarter of a century ago that highlighted the degree of unpaid community work in such neighbourhoods (e.g. Young and Wilmott 1975). On the other hand, there is a newer narrative of 'sink' estates. In this view, there is acute crime and fear of crime, the loss of community spirit and a sense of decline so far as quality of life is concerned as these populations become locked into increasingly debilitated and alienated communities (e.g. Social Exclusion Unit 1998).

Superficially, this study appears to support the view that there is greater solidarity in lower- than in higher-income areas. It finds that unpaid community work was used to perform a larger proportion of all tasks in

Table 5.3 Sources of unpaid community exchange: by area-type

	Lower-income areas	Higher-income areas
Kin	73	60
Non-kin	24	30
Groups	3	10

lower-income areas, especially in the urban context (see Table 5.1). However, before drawing this conclusion, it is first necessary to start to unpack both the extent and character of unpaid community work in these areas as well as the motives underlying participation.

Breaking down the 5 per cent of all the work undertaken on this basis, the finding is that in lower-income areas, some 73 per cent of these tasks are provided by kin, 24 per cent by friends or neighbours and 3 per cent by voluntary organisations (see Table 5.3). Such a finding is important because it reveals not only that nearly three-quarters of unpaid community work in lower-income areas is composed of kinship exchange but also that the provision of help by voluntary organisations is only a minor element in community exchange. This is replicated across all the study areas. In higher-income areas, for example, provision by kin constitutes again 60 per cent of all unpaid community work. The main difference is that a higher proportion is sourced from community-based organisations in higher-income localities.

Put another way, lower-income areas rely more on 'informal' rather than 'formal' unpaid help. By informal unpaid help we mean instances where people do an activity to benefit one or more others beyond the household/ family but do so outside any group context. Formal unpaid exchange, meanwhile, is where help is provided in a group context as members of some voluntary group or service volunteer programme. Until now, most studies have omitted to study informal aid-giving, instead focusing upon solely formal helping activity (Putnam 2000). The result is that they have underestimated the extent to which such aid exists in lower-income populations. In order to provide a fuller understanding of unpaid community exchange, each type is here examined in turn, starting with kinship exchange and then exchange with friends and/or neighbours followed by more organised unpaid exchanges.

Kinship aid

So far as kin are concerned, most interviewees were more than willing to engage in unpaid community exchange. Such a form of unpaid aid was widely supported across all of the areas studied and was frequently referred to as 'done out of the kindness of our heart', 'we like to help out', 'there were

Table 5.4 Percentage of households with kin living in the area: by locality-type

	Lower-income areas	Higher-income areas
Grandparents	20	5
Parents	67	23
Brothers or sisters	67	32
Children	68	51
Uncles or aunts	55	15
Cousins	61	15

family reasons' or 'we did it out of love'. Indeed, the only factor constraining both the giving and receiving of kinship support was the fact that many households had few kin living in the area (see Table 5.4). This was particularly the case in higher-income areas.

The greater prevalence of unpaid exchange in lower- than in higher-income areas, therefore, is in large part due to kin being present in such areas, enabling kinship exchanges to occur. For these populations, unpaid kinship exchange was a key component of their coping practices As Paugam and Russell (2000: 257) state:

> When a large part of the population shares the same unfavourable social conditions, familial solidarity does not arise from a logic of compensation nor from the logic of emancipation – it becomes a collective fight against poverty. Reciprocity in the exchanges is then functional. In order to face adversity, everyone gives and gives back, therefore, everyone gives and receives. This is why we are more likely to find examples of lasting familial solidarity in these regions where unemployment and hardship are higher, because it is based on a reciprocity imposed by the need to resist collectively.

However, if most interviewees were willing to engage in unpaid kinship exchanges in order to help each other to cope, this was not the case when non-kinship exchange was examined.

Non-kinship exchange

The common assertion was that others outside the kinship network could look after themselves so far as they were concerned. Beyond kin, and unlike earlier studies (e.g. Young and Wilmott 1975), norms of reciprocity were heavily imbued with payment in all neighbourhood types. There was little desire to help others on an unpaid basis.

This reflects an 'inner–outer' logic in unpaid exchange relations and shows how the principle of non-equivalent, spontaneous balancing of needs cannot be generalised outside of the 'inner' sanctum of kin. Outside of this inner kinship space, there is indifference and even hostility towards helping others. In the outer circle of friends, neighbours and others, unpaid help was not seen as something that should be provided.

Indeed, unpaid non-kinship exchange occurred only where it was felt to be unacceptable, inappropriate or impossible to do anything different. This tended to occur in four circumstances. First, it prevailed when it was felt to be unacceptable to pay somebody (e.g. when they lent you a saw). Second, it predominated when it was felt inappropriate to pay them (e.g. when a colleague from work did you a favour). Third, it occurred when payment was impossible (e.g. when somebody refused to be paid because they wanted an unpaid favour from you at a later date). And fourth and finally, it prevailed when the social relations prevented payment (e.g. when the recipient could not afford to pay and thus had no choice but to offer a favour in return).

If at all feasible, however, interviewees avoided unpaid exchange. On the one hand, interviewees expressed great anxiety about owing people a favour if they accepted an unpaid offer of help from somebody. As one woman in a deprived urban neighbourhood put it,

> I usually like to pay people who do work for me, so if I need to, I can feel free . . . but a friend laid my carpet and wouldn't take money. I owe him now and I really hate that hanging over me.

On the other hand, they also expressed concerns that if they helped others, the favour would not be returned. As one of the UK respondents in a lower-income urban area asserted,

> most people don't return favours these days so I don't do anything for anyone else unless I'm paid for it.

Remarkably similar findings were identified in a separate study in small town America (Nelson and Smith 1999: 112). As one of their respondents stated:

> We don't really like to owe anybody anything including favors because they can always come back on you in a negative way. So, whenever things are done it's usually been an exchange for pay.

In consequence, the widespread preference was for money or gifts to be involved in one-to-one non-kinship transactions. This avoided any obligation 'hanging over you' to reciprocate favours but at the same time, the wheels were being oiled for the maintenance or creation of closer relations without being 'duty bound'. Seen in this light, the greater prevalence of unpaid exchange amongst lower-income populations is simply due to their inability to pay for help rendered. Accepting an unpaid favour from a friend or neighbour was a result of their lack of choice. It was not a choice borne out of the existence of trust and/or close social relations (cf. Putnam 2000). This perhaps explains why a far greater proportion of one-to-one help received by higher-income households is paid (79 per cent) than is the case in lower-income households (23 per cent).

Organised community-based groups

On the whole, however, policy initiatives to harness such community exchange have not so far focused upon one-to-one reciprocity. Based on the idea that community-based groups and associations are more 'mature' forms of community self-help than one-to-one exchanges (see Home Office 1999) emphasis has been put upon developing formal groups. Here, therefore, we investigate the implications of such an approach.

The key finding is that it is largely affluent groups who join and participate in community-based groups, associations and organisations. Although 21 per cent of all lower-income households (i.e. earning less than £250 gross per week) participate in one or more community groups, some 75 per cent of higher-income households do so. In major part, this lack of participation in community groups by lower-income populations is due to a perception that they are for people other than them. Just 16 per cent of lower-income households (compared with 82 per cent of higher-income households) felt that they could join or participate. As a young unemployed single parent woman in a lower-income rural area put it,

> There is a real problem finding affordable childcare, so I found out about a babysitting circle where I live but they are all like really posh people so I wouldn't be allowed into that.

Or as another young single parent in the same area put it,

> If you are a single parent and poor, you are an outsider but you can be a total stranger but have a four-wheel drive and a posh house, suddenly you are accepted.

For many lower-income households, therefore, the perception is that community-based groups are for others, not them. An employed man in a lower-income rural household summed up the perception of many others in the following way:

> The community centre and all the community groups are more for upper-class people, not us.

For those (largely higher-income households) participating in such groups, 94 per cent asserted that they benefited in social, rather than material, ways. Joining provided the opportunity for interaction with other members so that advice, information, a listening voice and knowledge could be sought. Sports clubs, rambling associations, tennis clubs, village history societies and the like are thus the by-product of attempts by the already affluent to create social structures through which they can build and maintain social networks and seek social support.

The result is that great care must be taken when interpreting studies of social support that only examine community-based groups. Putnam (2000: 191), for example, concludes that

> Employed people are more active civically and socially than those outside the paid labor force, and among workers, longer hours are often linked to more civic engagement, not less

and that

> people with lower incomes and those who feel financially strapped are much less engaged in all forms of social and community life than those who are better off.
>
> (Putnam 2000: 193)

This however, is a result of solely examining organised aid. If the lens is broadened to one-to-one reciprocity, then the findings are very different. As Hall (1997) highlights in the context of the UK, most of the increase in civic activity has happened among the more affluent strata. People from poorer backgrounds are more likely to draw upon one-to-one reciprocity. This is reinforced by McGlone *et al.* (1998), examining the British Social Attitudes survey. They find that these more formal sources of help hardly figure in the coping practices of lower-income groups.

In major part, this can be explained in terms of the perception amongst lower-income groups that these community-based associations are for people other than them. In other part, however, it is also due to a lack of desire on the part of these households to receive charity. Most households went to great lengths to point out that they were returning a favour or expected to return a favour in the future when discussing unpaid community work. As Kempson (1996) has identified in another context, this is because they did not want to be seen as receiving charity and avoid such one-way social relations at all costs. Here, therefore, grave doubts must be expressed about any reliance on voluntary associations when attempting to rebuild social support networks.

In sum, the old-fashioned idea that lower-income populations help each other out on an unpaid basis is correct. It is not correct, however, that this occurs out of choice. Beyond kin, most people in most circumstances avoid using unpaid help and only do so when they cannot do the job themselves, afford to pay somebody, or when the social relations involved militate against payment. The result is that initiatives to develop reciprocity will need to reflect these circumstances. For these populations, the norm was to pay friends and/or neighbours so as to avoid any souring of their relationship if a favour was not returned, reflecting the fact that gifts and/or money are lubricating unpaid community work in an environment where trust was lacking. Initiatives to harness unpaid help beyond kin cannot avoid this desire for people to receive some form of payment, at least when the provision of material help is provided. Nor can one rely on existing community-based groups to harness such support networks. As shown, these are largely sociability vehicles for more affluent groups. They do not on the whole deliver material support to lower-income populations.

PARTICIPATION IN PAID INFORMAL WORK

The final informal practice considered here is paid informal work. This is often seen to exemplify the more flexible profit-motivated exchange relations that have arisen under post-Fordist regimes of accumulation (Castells and Portes 1989, Leonard 1994, Portes 1994, Sassen 1989). Conceptualising it as a form of low-paid peripheral employment conducted by marginalised groups for unadulterated economic reasons (e.g. Castells and Portes 1989, Kesteloot and Meert 1999, Portes 1994), proponents of this 'marginality thesis' assign such work to marginalised populations. Here, however, such an economic reading of paid informal work will be shown to be a misrepresentation of its distribution and meaning, at least in the context of contemporary England (for an in-depth examination, see Williams and Windebank 2001a).

As with previous studies (e.g. Fortin *et al.* 1996, Leonard 1994, Pahl 1984, Renooy 1990), this study confirms the concentration of paid informal work in higher-income populations. Take, for example, urban England. The populations of lower-income urban neighbourhoods (78.2 per cent of the sample) supply just 62.5 per cent of the paid informal work, whilst residents of affluent suburbs (21.8 per cent of the sample) supply 37.5 per cent of this activity. Comparing lower- and higher-income urban areas, the average amount received for conducting a task on a paid informal basis was £90.24 compared with £1,665. Moreover, the average hourly informal wage rate was £3.40 in lower-income areas but £7.50 in higher-income areas, and the mean annual household income from paid informal work was £46.22 compared with £435.62 (see Williams and Windebank 2001a).

However, our findings do not concur with the current perception that such work is everywhere economically motivated. Instead, it finds that although such work is firmly embedded in unadulterated economic motives in higher-income areas, this is not the case in lower-income localities. In these areas, much paid informal exchange is conducted for and by close social relations for primarily social reasons or to help each other out in a way that avoids any connotation of charity, and it is conducted more by women than men. In higher-income areas, meanwhile, such exchange is conducted more by self-employed people and firms for profit, is primarily used as a cheaper alternative to formal firms and is undertaken more by men than women (for an in-depth examination, see Williams and Windebank 2001c).

In lower-income urban areas, less than a third (31.7 per cent) of all paid informal exchanges used firms or unknown self-employed individuals but such sources constituted 84 per cent in higher-income urban areas (see Table 5.5). Therefore, in lower-income neighbourhoods, paid informal exchanges mostly involve transactions conducted by friends, neighbours or relatives, whilst in higher-income neighbourhoods these are between anonymous buyers and sellers. This is explained by the motivations of purchasers. In some 80 per cent of circumstances where paid informal work was used in affluent suburbs, it was employed as a cheaper alternative to formal employment but this is the reason for just 18.1 per cent of this work in lower-income neighbourhoods. In nearly all of these cases, it is firms and/or self-employed people not known by the household who conducted the work. When closer social relations are involved, it is either carried out for social reasons or seen as an opportunity to give money to another person in a way that avoids any connotation of charity (see Kempson 1996).

Hence, just because paid informal work is poorly reimbursed in deprived neighbourhoods does not mean that it is a low-paid form of peripheral *employment*. Much is conducted for close social relations and, as will be shown below, the exchange relations are more akin to unpaid mutual aid amongst kin (i.e. the social relations of the private sphere) than to an employer-employee relationship (i.e. the social relations of the public sphere).

Social reasons tend to predominate when friends or neighbours (rather than relatives) are involved. Indeed, it is in these paid informal exchanges between friends and neighbours that one can see mutual aid that supposedly used to take place on an unpaid basis. For participants, these exchanges are conducted in order to cement or consolidate social relationships. The exchange of cash is seen as a necessary medium, especially when neighbours or friends are involved, because it prevents such relations from turning sour if and when somebody reneges on their commitments. Cash thus provides the oil to allow mutual aid in situations where trust is missing. Indeed, many of the respon-

Table 5.5 Character of suppliers of paid informal work in higher- and lower-income urban and rural areas

% of all paid informal work conducted by:	Rural areas		Southampton		Sheffield		Both cities	
	H[a]	L	H	L	H	L	H	L
Firm/unknown person	5.4	8.3	92.1	29.6	76.6	33.3	84.0	31.7
Friend/neighbour	56.4	52.4	1.3	29.7	19.8	28.7	10.6	29.1
Relative	38.0	34.5	6.6	28.5	0.7	20.6	3.7	24.2
Household member	0.4	4.8	0.0	11.6	2.8	17.4	1.4	15.0

a Where H = Higher-income and L = Lower-income.

dents, especially in the deprived neighbourhoods, could not remember such relationships ever being any different. This, therefore, is not evidence of the formalised monetisation of previously unpaid reciprocal exchange. It is evidence that cash exchange does not have to be profit making in orientation and can be grounded in alternative social relations. Put another way, monetary exchange relations do not necessarily have to be capitalist. And it is here in these deprived neighbourhoods that solid evidence is available of the existence of monetised market-based relations devoid of profit-motivated intentions and capitalist social relations.

Besides social reasons, another rationale for paid informal exchange is redistribution. This mostly occurs when kin are involved. In these instances, it was either children or a relation such as a brother, sister or parent who was paid, normally in order to give them much needed spending money when, for example, they were unemployed. Indeed, using kin to do tasks so as to give them money was the principal rationale behind 10.9 per cent of all paid informal work in the lower-income neighbourhoods. For these consumers, therefore, it was a way of giving money to a poorer relative in a way that avoided all connotations of charity, even if this was an intention underlying such exchange. This type of motivation, however, hardly existed in affluent suburbs, except when young children were involved who were paid by their parents to do a job.

It is not only between higher- and lower-income neighbourhoods, however, that the ways in which people participate in paid informal work and their reasons for doing so vary. It is also the case when Sheffield and Southampton are compared. In Sheffield, a much higher proportion of paid informal work is conducted through relatives, friends and neighbours than in Southampton. For example, in the higher-income Southampton suburb, 92.1 per cent of paid informal exchange was conducted using people previously unknown to the customer but only 76.6 per cent in the Sheffield suburb (see Table 5.5). The reason is that this work was more likely to be conducted for profit-motivated reasons in Southampton than in Sheffield. In part, this is due to the relative inability of Southampton households to pay formal labour and in part due to their greater preference to engage in monetary relations rather than either do-it-themselves or seek unpaid help.

In sum, paid informal exchanges are largely driven by non-market motivations in lower-income neighbourhoods (and Sheffield) but by the profit motive in affluent areas (and Southampton). As with self-provisioning, therefore, participation in paid informal work appears to have a varying logic in different types of area. Indeed, these contrasting local logics mean that, although the profit motive has deeply penetrated monetary relations in affluent areas, some 30 per cent of all the monetary exchanges (both formal and informal) studied in these lower-income areas were conducted for rationales beyond the profit motive.

The implications of this finding are potentially profound. It shows that even if there has been the penetration of monetary relations into every nook and cranny of social life (cf. Harvey 1989, Sayer 1997), these are by no means everywhere driven by the social relations of capitalism. Monetary exchange is not everywhere based on profit-motivated market relations. Indeed, this study of paid informal exchange, often taken as an exemplar of profit-motivated monetary exchange, displays how it is wholly feasible to have monetary transactions that are embedded in alternative social relations and motives. This study locates such work in the social as well as the economic relations of poor communities. These transactions help maintain the social fabric of neighbourhoods and at the same time, fill the gap left by the failure of markets in goods and services. As such, we can only agree with Zelizer (1994: 215) that 'Money has not become the free, neutral and dangerous destroyer of social relations.'

In consequence, the task for policy is perhaps not to eradicate such work but to identify coping mechanisms that resonate with the motivations and attitudes towards community exchange in such areas (see Williams 2001b). Given that many are wary about engaging in informal exchange unless payment is involved, any policy initiatives to harness informal exchange will need to be based on some form of tally system or payment to conform to the norms of reciprocity in these populations. Initiatives that could fulfil these roles are the subject of Chapter 8 of this book.

CONCLUSIONS

In sum, this chapter has shown that informal modes of production are a key component of household work practices. However, at present, these informal practices reinforce rather than reduce the socio-spatial disparities produced by formal employment. Lower-income populations are less able to participate in informal work than higher-income populations and often engage in only a narrow range of activity that is conducted out of necessity. Higher-income populations, meanwhile, not only conduct a wider array of activity but more often engage in informal practices as a matter of choice.

If lower-income populations are to be enabled to draw upon the informal sphere as a coping practice, therefore, then the constraints that currently impede them from doing so will need to be tackled. The identification of these barriers is the subject of the next chapter.

6   Developing household coping capabilities
Problems and prospects

In the previous two chapters, it has been outlined how household coping capabilities and practices markedly differ socio-spatially. Lower-income populations are not only less capable of conducting work that they deem necessary but also engage in less informal work than their more affluent counterparts. In this chapter, the barriers that prevent people, especially lower-income populations, from participating in the informal sphere to enhance their coping capacities are identified. This will reveal that households are hindered from engaging in informal modes of production by their lack of human, social network, financial and time capital as well as environmental and institutional constraints. Unless these are overcome, then households will be unable to draw upon the informal sphere to improve their coping capabilities.

Given this formidable range of barriers to participation in informal work confronted by lower-income populations, some might believe that it would be easier for policy to confine itself to the 'employment-focused' approach towards poverty alleviation of conventional third way thinking. Rather than harness the ability of people to participate in informal work to enhance their coping capabilities, it might be assumed that pursuing social integration through insertion into employment alone is the more suitable option. In the second half of this chapter, in consequence, the implications of insertion into employment, especially low-paid employment, on capabilities is analysed. So too are the implications of households adopting informal-orientated rather than commodified practices. This will reveal that while inserting people into low-paid jobs has little overall positive benefit on the coping capabilities of households, encouraging households to pursue informal-orientated practices has the potential to significantly improve overall capacities.

BARRIERS TO PARTICIPATION IN INFORMAL ECONOMIC ACTIVITY

Evaluating the coping capabilities of households and their practices is an important first step in challenging many of the myths that have grown up surrounding poverty, especially with regard to the ability of the poor to fend for themselves. However, merely challenging myths cannot and should not be an end in itself. As Chapter 1 highlighted, it is unlikely that sufficient jobs can be created to allow everybody to use employment as a route out of poverty. At the same time, there appears to be a shift of economic activity from the formal to the informal sphere as highlighted in Chapter 2. Yet as shown in the last chapter, lower-income populations are less able than higher-income households to engage in this informal sphere to offset the disadvantages that they confront in the formal sphere. The outcome is that informal work currently consolidates, rather than diminishes, the inequalities produced by employment.

A crucial question is thus whether informal work can be harnessed in order to provide households and communities with additional means of meeting their needs and wants (see, for instance, Burns and Taylor 1998, Donnison 1998, Macfarlane 1996). To develop initiatives for enabling populations to develop their capacities to help themselves to a greater extent than at present, a first step is to understand the barriers to participation in informal economic activity. Here, in consequence, evidence from both previous studies and the English localities survey is used to set out the principal barriers to participation in the informal sphere.

In the context of the English locality studies, these barriers have been identified in several ways. First, when households had not undertaken a task either for themselves or others, they were asked why not. Second, and using Likert scales, households were asked to agree or disagree (using a 5-point scale from strongly agreeing to strongly disagreeing) with a range of statements concerning the constraints preventing them from engaging in more self-provisioning, unpaid community work and paid informal work. These explored the principal constraints identified in previous studies of informal work (e.g. Pahl 1984, Renooy 1990). In the case of self-provisioning, they included whether the provision of more time, money, skills, equipment and a different living environment would lead them to engage in greater amounts of self-provisioning and whether more money would mean that they would do less of such work. So far as paid informal work is concerned, meanwhile, they were asked whether time, opportunity, skills, equipment and fear of being caught were constraints on their doing more of such work and whether they would do less if they had more money and were offered a proper job. Finally, households were also asked in an open-ended manner what would lead them to engage in more self-provisioning, unpaid community work and paid informal work. The results, along with those of previous studies, are reported below.

The finding is that although most households (98 per cent) wish to engage in more informal work, six key barriers prevent them from doing so. These barriers incidentally, conform to the range of constraints identified in the report of the Task Force on Community Self-Help set up under the auspices of the Social Exclusion Unit (Home Office 1999). They cannot engage in greater levels of informal work principally due to their lack of social network capital, time, money, skills, physical ability and equipment as well as institutional and environmental barriers. The suggestion, therefore, is that if these barriers to participation could be addressed, then there would be an opportunity for the growth of such activity. Here, we review each barrier in turn.

Social network capital

A first barrier to participation, especially so far as the giving and receiving of paid and unpaid support is concerned, is that some 58 per cent stated that there was little opportunity to do so. Exploring why this is the case, our finding was that although people might know a wide number of people in their communities, they do not feel that they know them well enough to either ask or be asked to do something.

Having identified this lack of social network capital as a key barrier to participation in informal work during the 1998–9 survey of urban areas, the decision was taken in the rural survey conducted the next year to include some additional questions to further explore this issue. Here, the results are reported. This examined whether rural households could name their next-door neighbour and whether they knew many people in the street or immediate area. Previous studies view rural areas to have thicker social networks than urban areas (e.g. Chapman *et al.* 1998, Cloke *et al.* 1994, Findlay *et al.* 1999, Hedges 1999, Stratford and Christie 2000). This study was little different. The finding was that they nearly all knew their next-door neighbour, although this is slightly lower in the affluent rural areas of Fulbourn and Chalford (see Table 6.1) and many knew the people in their

Table 6.1 Would you be able to name your next-door neighbour?

Immediate neighbours	Grimethorpe	St Blazey	Chalford	Wigton	Fulbourn
Know them	94.3	95.7	92.9	97.1	88.6
Don't know them	0	2.9	7.1	2.9	11.4
Not applicable	5.7	1.4	0	0	0

Table 6.2 Do you know many people in the street/immediate area?

Knowledge of households in the street	Grimethorpe	St Blazey	Chalford	Wigton	Fulbourn
Hardly anybody	15.7	5.7	8.6	8.6	34.3
Several	4.3	18.6	7.1	12.9	25.4
Half of them	21.4	21.4	18.6	22.9	15.7
Most of them	17.1	17.1	48.6	32.9	14.3
All of them	41.5	37.1	17.1	22.9	10.0

Table 6.3 Number of people named as close friends (not including family): by area

	Grime-thorpe	St Blazey	Chalford	Wigton	Fulbourn
Close friends in the community					
0	2.9	5.7	4.3	7.1	15.7
1–3	15.7	25.7	27.1	27.1	37.1
4–7	31.4	32.9	35.7	27.1	18.6
8+	50.0	35.7	32.9	38.6	28.6
Amount of contact					
< Once a year	7.1	0	1.4	1.4	2.9
Special occasions	0	0	1.4	1.4	0
Only when needed	4.4	4.3	0	1.4	5.7
Occasionally	20.0	4.3	18.6	5.8	4.3
Regularly	68.6	91.4	78.6	90.0	87.1

street/immediate area, although this was again lower in affluent rural communities (see Table 6.2).

Based on this, one might think that there are many opportunities for these rural populations to engage in informal exchange. Indeed, this is reinforced when the number of close friends in the community and the degree of contact they have with them are investigated. Most households have a number of 'close' friends in the area and maintain regular contact with them (see Table 6.3). However, questions on why they do not help out friends and neighbours on an unpaid basis reveal a widespread perception that they do not know people well enough to either ask or be asked to do something. Instead, the common belief is that it is not 'the done thing' to request and offer help to others. As a male aged 65+ years old put it, after telling us how everybody knew each other in his apartment complex,

> Even in the block of flats where I live, you wouldn't call on people if you needed help, well I suppose you would if you were desperate, but not as a rule.

There is thus a common reticence about asking for help or offering help to others. Typical of this was the experience of a woman aged 18–24 who lived in the deprived rural locality of St Blazey,

> There is not that much like, real community spirit, not genuine people wanting to help. For example on our estate we've got goodness knows how many kids, I've lost count there's so many, but most of them go to the local school. Five or six of them go in different cars with one child in each and no way would one of them stop to give a lift to the people that were walking. Even if it's absolutely blagging it down with rain, you just get soaking wet and they just drive on past. What's the point

in that. I've lived there for three and a half years and in the last few months I've only just started asking my neighbour to give me a lift. He's driven past me so many times during the last few years and it's only now that I've plucked up courage to ask him for a lift and it's only because I didn't want to burden him.

As outlined in the last chapter, the widespread perception was that if you offered to help somebody, they would think that you saw them as a 'charity case' and this would sour your relationship with them.

Even when these rural populations decided to ask for help or to offer it, however, the circle of people that they felt able to call upon was very much delimited by socio-economic context. Take, for example, jobless working-age rural households. These households perceived themselves to have very 'thin' and 'narrow' social networks to which they can turn. First, this is because of their inability to repay a favour. Many no-earner households, that is, include the physically disabled or retired. Second, it is because of the smaller size of social networks that results from being unemployed (Engbersen *et al.* 1993, Kempson 1996, Morris 1994, Renooy 1990, Thomas 1992). The long-term unemployed in particular mix mostly with other long-term unemployed. They also have relatively few friends or acquaintances in employment (Kempson 1996, Morris 1994) and the majority of their unpaid community exchange is between friends and acquaintances (Kempson 1996, Van Eck and Kazemieer 1985). They are also less likely to have kin living locally. The result is that jobless households have fewer people to call upon for aid than households with people in employment. Third and finally, even where kin or employed acquaintances are present, the widespread perception was that one would not ask them for help because they would perceive you as 'on the take' (male Grimethorpe resident).

Another way in which socio-economic context constrains whom one can call upon for help arose during focus group discussions with young parents in St Blazey, Cornwall. It is not simply that one's social networks may be short or not very dense, but there is a widespread perception that class, gender and/or age result in exclusion from certain networks. This was apparent when talking to young parents who felt that their age, class and in some cases, gender, excluded them from some existing networks in their area that would be useful to them. As a male homemaker and principal childcarer in St Blazey put it,

When the baby was born my partner carried on working and I stayed home to look after the baby and when I used to take the baby to play-group and that, they were all mums and they all used to look down on me and treat me as if I was something different. They would all go off and have tea and biscuits and I would never be invited and because of that my son would miss out on the interaction with other children. Wherever we go it's like a closed network, it's still the same.

Similar feelings of exclusion, this time in terms of 'class', were witnessed by young mothers in St Blazey. As a woman aged 18–24 years old expressed it,

> There is a real problem finding affordable childcare, so I found out about a babysitting circle where I live but they are all like really posh people so I wouldn't be allowed into that.

Or as another young single parent put it more explicitly,

> If you are a single parent and poor, you are an outsider but you can be a total stranger but have a four wheel drive and a posh house, suddenly you are accepted.

In sum, a common barrier amongst these rural people is that they do not feel that they know people well enough to ask or be asked to do something, or they feel excluded on the grounds of age, gender and socio-economic status from joining community-based groups.

Hence, if lower-income populations in general and jobless households in particular are to engage in greater levels of informal work, then there will be a need to widen their social networks and thus social support structures. This is increasingly recognised in the social capital literature (e.g. Coleman 1988, Putnam 1993, 1995a, 1995b, 2000). This argues that the 'strong' ties commonly but not always associated with kinship and close-knit communities may actually be less effective than a large and more diverse network of ties that are developed through other social networks. Thus, Granovetter (1973) writes of the 'strength of weak ties'. Weaker ties might have limits on the claims that can be made on them, but they also tend to provide indirect access to a greater diversity of resources than do stronger more socially homogeneous ties. As Granovetter (1973: 1,371) points out, 'those to whom we are weakly tied are more likely to move in circles different from our own and will thus have access to information different from that which we receive'. Put another way, diversity represents strength because it provides access to a wider variety of opportunities and perspectives on issues and problems, an idea also developed by Perri 6 (1997).

The problem, however, as Burns and Taylor (1998) point out, is that the reality for many lower-income populations is that they have neither the dense overlapping networks of yesteryear nor the sparse overlapping networks required in today's world. In consequence, if informal work is to be facilitated, then both these strong and weak ties will need to be further developed. How this can be achieved will be addressed in Part III. Tackling the barrier of social network capital, however, is a necessary but insufficient means of facilitating participation in informal economic activity. There is little use having large social networks if one has little to offer or lacks the time to maintain and use them.

Time capital

For more than half of all households (52 per cent), lack of time is a constraint on participation in informal work. This applies to their ability to engage in both self-provisioning and informal exchange. Such a constraint is most acutely felt among those spending long hours in their formal employment and/or those who commute long distances in relation to their job. Some 76 per cent of all multiple-earner households defined time as a barrier to their participation in informal economic activity. Many of these households wanted to do jobs for themselves but were unable to do so due to their long hours in the employment-place. When sufficiently paid to be able to pay formal labour to do these tasks, this is not a problem so far as the coping capabilities of households are concerned. However, when multiple-earner households receive relatively lower incomes, then the impacts of long hours can be acutely felt in terms of their ability to get necessary work completed. As one multiple-earner household living in a lower-income urban neighbourhood put it,

> We both work long hours just to keep our jobs and when we get home we are really knackered. You don't feel like doing anything. The problem though is that at least if you are unemployed, or one of you isn't working, you get the time to do things. We don't. And we definitely cannot afford to pay the prices builders and everyone charges. So loads of jobs that need doing get left undone. It's not just repairing the broken gutter. It's loads of things. We often wonder whether it's worth us working.

Another impact of long hours spent in employment is that many households complain that their social networks tend to be largely based on the employment-place. This unravels yet another facet of the 'work/life balance' problem. It reveals that it is not simply the case that the time spent in employment is reducing the time available to spend with children and kin; it is also reducing the wider networks of social and material support beyond the employment-place. The outcome is that their struggle to earn sufficient to make ends meet often results in a contraction of their wider support networks, spiralling them yet further down the path of needing more money in order to pay for formal services.

Some 89.4 per cent of multiple-earner households in deprived urban neighbourhoods, as a result, asserted that they would engage in greater levels of informal economic activity if they had more time. Indeed, the complex problems confronted by these households provide many lessons for the current policy approach of employment-led social inclusion. As pioneers of the 'employment is the route out of poverty' model these households provide a useful exemplar of its problems. Earning low pay yet working long hours, these households found themselves not only unable to maintain social networks outside of the employment-place but also without the time or energy to engage in self-provisioning activity and informal exchange. The result was that they had to rely even more heavily on formal labour, which they could ill afford, in order to get necessary tasks completed around the home due to their lack of time.

Human capital

Besides social network and time barriers to participation in informal work, there are also human capital constraints (see Fortin *et al.* 1996, Howe 1988, Renooy 1990, Smith 1986). These take two forms. First, people perceive themselves to lack the skills necessary to help out others. Indeed, 52 per cent of households agreed that they would engage in more informal work if they had more or different skills. Second, there is the issue of health. Some 28.2 per cent asserted that their health prevented them from doing more for others and this rose to 40 per cent in the deprived ex-pit village of Grimethorpe.

This reveals therefore that skills and health are necessary not only for insertion into the formal labour market but also for informal modes of production. For example, many asserted that they would have liked to help out others with maintaining their home (e.g. decorating, mending broken windows, repairing guttering, gardening) but felt that they did not have the skills to do so. As one male in the lower-income rural locality of Wigton put it, 'How could I help others. I haven't got the skills. I would just mess their home up.' The human capital constraint, however, is not simply one of skills. It is also one of a lack of confidence. Many who are unemployed quite simply lack the confidence to offer to help out others. As many jobless households responded when asked about doing work for other households, 'I don't have anything to offer', 'what could I do' and 'they wouldn't want me messing up their house'.

More importantly, poor health prevents over a quarter of households from giving or receiving help in the communities studied. On the one hand, this means that they are unable to contribute to helping others. On the other hand, and perhaps more saliently, it means that they feel unable to ask others for help because they feel unable to reciprocate the favour. Indeed, it is in these households that help tended to be rewarded with gifts and/or money to a greater extent than in other households. For them, such payments acted as a substitute for reciprocity and prevented social relations turning sour when they were unable to repay favours.

This human capital barrier is relevant across the whole spectrum of informal economic activity. Take, for example, paid informal work. Having a formal job often means that customers recognise a person as having a skill to offer and it is a legitimisation of their skills in the eyes of the recipient. As many asserted, they used a particular person to get a job done (either paid or unpaid) because 'it is their trade' or 'they have the skills because they do it for their job'. Those without formal jobs thus suffer in terms of gaining access to paid informal work. There is no legitimisation of their skills. It is not solely a perception of skills, however, that prevents the unemployed from gaining access to paid informal work. If their skills are inappropriate for finding formal employment, there seems little reason to believe that they can sell or exchange them on the informal labour market. Indeed, perhaps this helps explain why multiple-earner households engage in more paid informal work than no-earner households. Having a formal job means that the outside world recognises a person as having a skill to offer and is a legitimisation of these skills in the eyes of potential customers. It is also because these households quite simply have skills that are desired by customers.

Economic capital

Besides social network, time and human capital, a further barrier to participation in informal economic activity, especially amongst lower-income populations, is that they often lack the money to acquire the goods and resources necessary to engage in such work. This barrier has been identified elsewhere (e.g. Pahl 1984, Smith 1986, Thomas 1992). The result of having little or no disposable income is that a household's access to many of the ways in which it could help itself is curtailed. For example, if one cannot afford the paint, brushes and sandpaper, then one cannot paint exterior windows so as to prevent the degradation of the fabric of the household.

Overall, some 52 per cent of all households asserted that they would engage in greater amounts of self-provisioning if they had more money. Such an explanation, however, was particularly prominent amongst jobless households. Some 70 per cent of no-earner households perceived money as a barrier to their participation in self-provisioning activity compared with just 43.2 per cent of multiple-earner households.

Interestingly, it was not the case that if households had more money, they would do less for themselves (i.e. the formalisation thesis). Only 3 per cent of households asserted that they would engage in less self-provisioning if they had more money. The vast majority said that it would enable them to do more. As such, there is a popular perception amongst people that they want to engage in more informal economic activity. Money is a constraint that prevents this being achieved and if they had more money, their perception was that they would do more for themselves, not less.

A direct product of having insufficient money was that households could not access the equipment to engage in self-provisioning and informal exchange. As a result, many perceived a lack of equipment as a principal constraint on their ability to engage in such activity. Some 48 per cent of households asserted that they would do more self-provisioning and 43 per cent more paid informal work if they had the right equipment. For example, when asked if there is any work that they have not done that they would like to complete, households often cited activities like redecorating, installing a shower, creating a workshop or even hoovering. In nearly all cases, they could not do these activities due to their lack of money and/or their lack of tools.

For instance, without access to a car, it is frequently more difficult to engage in self-provisioning. Take, for example, engagement in DIY activity. If there is bus access to DIY stores and one manages to overcome this hurdle, it is then necessary to carry home the ladder, pots of paint, wallpaper or other goods on the bus. This is not an easy task, as anybody will testify who has ever tried to do it. Moreover, without a car, it is also often more difficult to travel to engage in unpaid or paid informal exchange. The outcome is that without access to a car, one is unlikely to be able to easily engage in both self-provisioning and unpaid or paid informal exchange for others. This has a knock-on effect in the sense that others are then less likely to reciprocate favours. Indeed, examining the results from the survey of affluent and deprived neighbourhoods, those no-earner households without access to a car received considerably less paid and unpaid informal exchange than those no-earner households with access to a car, displaying the extent to which ease of mobility matters.

It is not only a car that is vital to engage in self-provisioning and informal exchange. Those lacking economic capital also cannot gain access to many other vital pieces of equipment with much lower unit costs in order to engage in such activity. If households do not possess tools such as a ladder and workbenches, they cannot conduct many tasks that are essential to maintain or prevent the degradation of the fabric of their dwelling (e.g. cleaning gutters). In consequence, money matters. Without it, the opportunities to engage in informal work are severely limited.

Institutional barriers

Another barrier to participation in informal economic activity, particularly for benefit claimants, is that they feel more inhibited for fear of being reported to the authorities and having their benefit curtailed. This was cited as a barrier to engagement in such work by 5 per cent of all respondents. Superficially, therefore, it does not appear a major barrier to participation. However, the fact that 57 per cent of the registered unemployed cited this displays that for this group, it is a major constraint. For example, an unemployed member of a no-earner household possessed the desire to set up a facility in their home to commercially utilise their skills in glass painting and textile making but had not done so due to what they referred to as 'the social security trap'. That is, if they earned over the 'income disregard limit' their benefit allowances would be reduced pound for pound. Such households also feared that neighbours might report them to the authorities. Indeed, such fears are perhaps not without foundation given the way in which working informally whilst claiming social security is considered to be a more serious offence than engaging in tax fraud (Cook 1997, Dean and Melrose 1996, Jordan *et al.* 1992). Reflecting this, many respondents in our survey expressed vehement opposition to people being paid informally, especially the unemployed. Their fear is thus perhaps well founded. Some unemployed respondents even expressed a fear of engaging in unpaid community work in case they were mistakenly reported to the authorities.

At present, the UK policy approach towards paid informal work is similar to that of international organisations (e.g. European Commission 1998a, ILO 1996, OECD 1994) and other national governments (e.g. Hasseldine and Zhuhong 1999). It is to eradicate paid informal work through a stringent environment of deterrence based on increasing the likelihood of detection and prosecution (see Grabiner 2000). This assumes that participants in paid informal work are rational economic actors. It then seeks to deter them from engaging in such economically motivated work by ensuring that the expected cost of being caught and punished is greater than the economic benefit of participating in such activity (e.g. Allingham and Sandmo 1972, Falkinger 1988, Hasseldine and Zhuhong 1999, Milliron and Toy 1988). The reason for seeking its eradication, meanwhile, is that it is seen to both represent unfair competition for formal activity and thus have a deleterious effect on formal employment, and undermine the welfare state by depriving it of income that could be used for social cohesion purposes (Williams 2001b). Put another way, such work is seen to disrupt the smooth and efficient running of formal economic and welfare systems (Feige 1979, 1990, Grabiner 2000, Gutmann 1978).

Until now, the principal issue investigated in relation to this deterrence approach has been whether it can be effective (e.g. Alm 1991, Hasseldine and Zhuhong 1999, Weigel *et al.* 1987). Little regard has been paid to the consequences of such an approach for everyday life. This is perhaps because it is perceived that greater insertion into employment, resulting from the jobs created by its eradication and the savings on benefit fraud, will provide current users with access to income to purchase formal substitutes. This, however, assumes that paid informal work is simply used as a means of making or saving money and that the task is to divert such activity towards formal mechanisms that can achieve these ends. In Chapter 5, nevertheless, it was revealed that only a small proportion of all paid informal work is undertaken in order to make or save money. Instead, the vast majority is akin to unpaid community exchange in that it is conducted in order to help out others and/or to cement and build social networks.

There is thus little evidence that its eradication will result in a growth of formal jobs. As outlined in Chapter 5, users would instead mostly do the work themselves if they did not informally pay these people. Indeed, Mogensen (1985) finds much the same in Denmark, identifying that two-thirds of such work in Denmark would not be converted into formal employment. Consequently, it cannot be assumed that its eradication will lead to formal job creation to provide people with access to money so that they can afford to purchase substitutes.

A further issue is whether it is appropriate to attempt to eradicate such work without putting in place alternative mechanisms for the provision of informal support. Paid informal work currently acts as a form of social support in these neighbourhoods used both to help out others and to cement and build social networks. Eradication of such activity without putting in place substitutes would thus serve merely to reduce the level of community exchange yet further. It is already the case that a lack of trust necessitates that such exchange requires cash as a medium. To increase the deterrents attached to engaging in such work without putting in place alternatives is thus likely to diminish yet further the little community exchange that currently manages to survive in such neighbourhoods.

In consequence, the task is to identify coping mechanisms that not only provide a substitute to paid informal work but which also resonate with the motivations and attitudes towards community exchange in such localities. In these lower-income areas, to repeat, many are wary about engaging in informal exchange unless payment is involved. As such, any alternative will need to be based on some form of tally system or payment to conform to the norms of reciprocity in these populations. Initiatives that could fulfil these roles will be returned to in Part III. For the moment, there is a need to explore the other barriers to participation in informal work.

Environmental barriers

Another barrier to participation in informal work is the type of area in which people live. So far as unpaid and paid informal exchange is concerned, many adopted the attitude of 'keeping themselves to themselves' in lower-income areas due to a perceived lack of trust, community and sense of well-being around them. Respondents wished to engage in closer social relations with others but had taken on board the image constructed of their area by outsiders that it was a dangerous place. This negated their desires to get to know other local people. Indeed, many asserted that they would engage in more community exchange if they lived somewhere else.

The perception of the area resulted in a low level of unpaid and paid informal support. However, it was not only the perception of the locality that led to a low level of such exchange. The lack of money in such areas also results in diminished levels of informal exchange. Take, for example, the task of window cleaning. Most households in these deprived areas do this task themselves. There is little point, therefore, being a window cleaner in this area if few employ people to have their windows cleaned. As such, there is simply less demand for informal exchange in these areas than in other more affluent localities. The important point here is that it is not only the lack of money in these localities that leads to a diminished level of informal activity. It is also the sense of community. Inhabitants often take on board the image of their own area that is depicted in media and local government narratives and this leads them to further implode on themselves, resulting in even less sense of community and social support.

In sum, the widespread desire to engage in greater amounts of informal work is curtailed by numerous barriers including time, money, social networks, skills and physical ability as well as institutional and environmental barriers. If these barriers to participation in informal work could be addressed then there would be an opportunity for the growth of such activity. In order to tackle these barriers, however, recognition is required of the prevailing attitudes towards informal work in contemporary England. It is already the case that a lack of trust necessitates that informal exchange requires cash as a medium when it is conducted beyond immediate kinship networks. Given that many are wary about engaging in informal exchange unless payment is involved, any policy initiatives to harness such exchange will need to be based on some form of tally system or payment to conform to the norms of reciprocity in these populations. Initiatives that could fulfil these roles are the subject of the next section of this book. Before doing so, however, we conclude this section by examining some further reasons why it is necessary to harness informal economic activity. To do this, we explore the implications of increased participation in first, the formal sphere and second, the informal sphere, on household coping capabilities.

IMPLICATIONS OF INCREASED PARTICIPATION IN THE FORMAL SPHERE ON HOUSEHOLD COPING CAPABILITIES

A key problem when the members of a household engage in employment is that they have less time to devote to maintaining and developing their local social networks and less time to spend on self-provisioning. The result is that households then have to source a wider range of formal goods and services, resulting in them having to spend their income on goods and services that they before perhaps provided on an informal basis. Employment thus leads to a formalisation of household coping practices to provide the same goods and services as previously. For higher-income households, this is not a problem at least so far as overall coping capabilities are concerned. The only problem was that some talked of feeling like a 'hamster on a treadmill' or 'stuck in the rat race' when describing how they earned money to pay people to do jobs that they could do if they did not have to spend so much time in employment.

Is it the same, however, for lower-income households? Does participation in employment enhance their coping capabilities? The potential problem is that the income that they receive may be less than what is required to compensate for what they could otherwise source on an informal basis. If this situation arises, then the net outcome is that entry into the formal labour market will result in a reduced level of coping capability than would otherwise be the case.

To investigate whether this occurs, the coping capabilities of lower-income multiple-earner households (earning less than £250 per week) compared with those of jobless working-age households are explored. Although multiple-earner households as a whole have much higher coping capabilities than jobless households, the coping capabilities of lower-income multiple-earner households are less than those of jobless working-age households. Whereas lower-income multiple-earner households got 51.1 per cent of the tasks completed that they wished to accomplish, jobless working-age households managed to complete 56.7 per cent of these tasks. The implication, therefore, is that as a household, one is better off remaining jobless than taking multiple low-income jobs, at least measured in terms of one's ability to get necessary work completed.

To explain this, the coping practices adopted by these two household types need to be examined. Low-income multiple-earner households adopt much more commodified work practices than their jobless counterparts, reinforcing the above assertion that their lack of time means that they have to formalise a greater share of their domestic provisioning. They thus engage in far lower levels of self-provisioning and more often cite a lack of time and a lack of money as their reason for not conducting tasks that they perceive as necessary. Their social networks, moreover, appear to be thinner. Jobless working-age households stated that they had a larger average number of close friends than lower-income multiple earner households (4.5 compared with 2.9) and were twice as likely to see them at least once a week. Furthermore, while 65 per cent of jobless working-age households asserted that they were unable to draw upon the help of others, this figure was 73 per cent in lower-income multiple-earner households.

The key question to be answered, therefore, is whether there is a level of income at which the coping capabilities of multiple-earner households excel those of jobless working-age households. If this can be found, then the household income level at which 'employment pays' and it becomes disadvantageous to be jobless will be known. Our finding is that this is spatially variable. This is due to the geographically variable interrelationship between cost of living, social payments and wage levels discussed in Chapter 4 that results in spatial variations in the extent to which 'employment pays'.

In Southampton, employed households need to earn £14,300 per annum before their coping capabilities outstrip those of jobless households, whilst in Sheffield this figure is slightly lower at £13,000; multiple-earner households need to earn £15,600 and £14,300 respectively. Although this does not take into account different household compositions (e.g. households with children, single-parent households and so forth), it does provide a broad measure of the basic guaranteed minimum wage required in these two cities in order to 'make employment pay'.

Since October 2002, the Working Families Tax Credit has been set at £237 per week (or £12,324 per annum), well below the rate required to make employment pay in terms of improved household coping capabilities. This assertion, however, needs to be treated with caution. On the one hand, we requested gross household income in only £25 per week (or £1,300 per annum) intervals, so the figures stated of required annual incomes are broad. On the other hand, these figures do not compare households of similar types. Instead, they put together disparate households with different compositions.

In the sense that different household types have varying income requirements to achieve the same standard of living, it is thus erroneous to put together households with contrasting compositions. To equivalise income, two techniques are available. The first is the McClements equivalence scale, which is the standard measure used by the Office for National Statistics (Government Statistical Service 1998). The second is the measure used by Gordon *et al.* (2000). They suggest that the 'head of household' (*sic!*) should receive an equivalence value of 0.70, a partner 0.30, each additional adult (over 16) 0.45, the first child 0.35 and each additional child 0.30 and if the head of household is a lone parent, 0.10 should be added. Hence the equivalence value for a household composed of a couple with two children is 0.7 + 0.3 + 0.35 + 0.3 = 1.65 compared with a figure of 1.0 for a couple without children. If both households have a gross household income of £10,000, then the equivalised income for the couple with children is £6,060 (i.e. £10,000/1.65). For a single person living alone, in contrast a gross household income of £10,000 would have an equivalised income of £14,285 (£10,000/0.7). The suggestion, therefore, is that a single person living alone is far better off on a gross household income of £10,000 than a couple with two children.

Given these strong differentials between household-types regarding the level of income required to provide the same standard of living, the decision was taken to compare the household coping capabilities of multiple-earner and jobless households for an equivalent household type, namely couple households with two children. The finding is that couple households with two children where both are in employment need to earn about £16,900 per annum in Southampton and £15,600 in Sheffield before they exceed the household coping capabilities of their jobless working-age counterparts with two children. It is only at these levels of gross household income that their household coping capabilities start to outstrip those of their jobless counterparts. The message, therefore, is that these households need a guaranteed minimum income of at least £325 and £300 per week respectively in order to make employment pay. Put simply, the level at which the WFTC is set is insufficient to 'make employment pay' in terms of improving household coping capabilities. WFTC presently does not take into account how entry into the formal labour market results in the formalisation of goods acquisition and services provision that may before have been undertaken on a self-provisioning or informal basis. In a nutshell, earning money costs money.

IMPLICATIONS OF INCREASED PARTICIPATION IN THE INFORMAL SPHERE ON HOUSEHOLD COPING CAPABILITIES

While the amount of money required to maintain coping capabilities increases when households enter the labour market, the amount required reduces when a household engages in greater levels of informal economic activity. Households of the same income bracket had remarkably different household coping capabilities measured in terms of their ability to get necessary work completed depending on whether they erred towards the formal market to service their needs or whether they were more likely to rely on the informal sphere.

To show this, all of the 861 households surveyed can be divided into two different types. Taking multiple-, single- and no-earner households in turn, we classified them into households that rely on commodified labour to a greater extent than the average household in their household type, or what we here call 'formal-orientated' households. Conversely, those relying less than average for their household type on commodity labour, we identified as 'informal-orientated' households.

Examining multiple-earner, single-earner and no-earner households in turn, 'formal-orientated' households were found to have lower overall coping capabilities than 'informal-orientated' households in every category. Take, for example, no-earner households. Formal-orientated jobless households managed to complete only 53 per cent of the tasks that they viewed as necessary whilst informal-orientated jobless households had managed to undertake 61 per cent of these necessary tasks. On the whole, those jobless households that tended to be formal-orientated were predominantly those with relatively few kin living in the locality, those who had a shorter length of residence in the area than informal-orientated households, mixed ethnicity households and those with lower formal educational qualifications.

To compare the impacts of a formal-orientated household and an informal-orientated household when there are multiple-earners, these households were divided into lower- and higher-income multiple-earner households. Starting with lower-income multiple-earner households (the 'working poor'), the formal-orientated working poor managed to complete 48 per cent of the tasks that they deemed necessary while the informal-orientated working poor undertook 55 per cent of such work. In consequence, the capabilities of the 'working poor' were higher when they pursued informal coping strategies rather than formal strategies. Working-poor households who tended to adopt informal orientated practices were mostly those with relatively thick kinship networks in the locality (upon whom they relied for unpaid and paid informal help) and a relatively long length of residence in the area. The working poor who tended to be more formal-orientated in their practices, meanwhile, were those with relatively thin kinship and non-kinship networks, including mixed ethnicity households who appeared to suffer more than most other households from being excluded from receiving outside informal support.

It was a similar pattern when it came to higher-income multiple-earner households. Formal-orientated multiple-earner households earning over £250 per week failed to complete 26 per cent of the work that they deemed necessary while this figure was just 18 per cent when such households were informal orientated in their practices. Amongst these multiple-earner households, those adopting informal orientated practices tended to have relatively strong kinship networks in the locality (upon whom they relied for informal aid), male household members employed in skilled manual occupations and relatively 'thick' social networks. Formal-orientated households were more usually composed of households with thin kinship networks, men in white-collar occupations and people who had lived in the locality for a relatively short period.

These results thus provide some important, albeit expected, findings. They show that households more able to draw upon outside sources of informal support and engaging in greater levels of self-provisioning have higher household coping capabilities than those unable to undertake such informal practices. The implication, therefore, is that policies that solely focus upon inserting citizens into the formal labour market do not necessarily improve the overall capabilities of households. Indeed, if they insert them into low-paid jobs, they reduce the overall coping capability of the household. The lesson, in consequence, is that if households can be encouraged to harness their ability to participate in the informal sphere, rather than rely solely on formal practices, then they will be able to enhance their coping capabilities to a greater extent than is currently the case.

CONCLUSIONS

Having shown in previous chapters how lower-income populations are less able to conduct work that they deem necessary and to draw upon the informal sphere in order to do so, this chapter has sought to explain why this is the case. To do this, we have analysed the various barriers that prevent lower-income populations from participating in the informal sphere in order to enhance their coping capabilities. This has revealed that such households are prevented from engaging in informal modes of production by their lack of human capital, social network capital, financial capital and time capital as well as environmental and institutional constraints. This provides us with the key hurdles that will need to be tackled if household coping capabilities are to be enhanced by developing the informal sphere and sets the scene for Part III, which investigates policies that can help households to overcome these barriers to participation.

Given that some might assert that these present a formidable range of barriers and that we should maintain an 'employment-focused' approach towards poverty alleviation, we have here explored the impacts on household coping capabilities of inserting people into formal employment as the best route out of poverty. This has revealed that many households that have entered the formal labour market but are confined to low wages have, if anything, lower household coping capabilities than jobless households. The lesson is that if employment is to represent a route out of poverty for jobless households, then the formal jobs need to provide a 'decent' wage. To put a figure on what represents a 'decent' wage, we have here estimated the level of income required by various household-types in order to have greater household coping capabilities than jobless households. Although these income levels are both spatially variable and depend upon the household-type being considered, the finding of this chapter is that they are in most cases above the level at which tax credits are currently set. Whether sufficient jobs can be created that are paid at the levels necessary to result in an improvement in household coping capabilities is very much open to question.

However, while inserting people into low-paid jobs has little overall positive benefit on household capabilities, encouraging households to pursue informal-orientated practices has been here shown to have the potential significantly to improve overall capabilities. Households that are informal-orientated have much higher overall capabilities than those formal-orientated in their practices. The lesson, in consequence, is that if households can be encouraged to harness their ability to participate in the informal sphere, rather than rely on formalised practices, then they will be able to improve their coping capabilities to a greater extent than is currently the case. In sum, this chapter reveals that to improve household coping capabilities, it is necessary to focus upon participation in the informal sphere. For this to occur, the barriers to participation identified above need to be addressed. How this can be achieved is the focus of Part III of this book.

Part III

Tackling poverty

A third way approach

7 Towards a 'civil-ised' society

From full-employment to 'full-engagement'

In recent years, there has been recognition that academic enquiry has often shunned discussing the implications of its research findings for policy-making. On the one hand, this is because there has been confusion on the part of some academics about what constitutes 'high-quality' research. A popular prejudice has arisen within the academic community that 'blue skies' research is high quality and that anything policy-orientated is 'applied' and of a generally lower quality. Examining policy issues has thus become tainted with negative connotations. The problem, however, is that there has been a failure to distinguish between repetitious policy-orientated consultancy projects that meet the routine needs of users, and research that explores fundamental principles and policy options in order to steer and enhance the policy community. The outcome is that policy-orientated research has been to a large extent avoided by many academics for fear that they will be tainted with a reputation for being 'lightweight' theoretically.

On the other hand, the emergence of post-modern discourses and the retreat from detailed empirical research, especially of a quantitative variety, has reinforced, rather than stemmed, the shift away from discussing policy issues. As one academic commentator puts it, the post-modern and cultural turns have led to forms of enquiry and writing that

> simply treat theory and concepts as a sort of intellectual game which has become increasingly detached from real world problems and concerns . . . Under the guise of liberation, empowerment and giving voice to those hitherto excluded, [this trend] simply reinforces the privileges of the intellectual elite to play an elaborate language game written by and for a tiny minority of participants.
>
> (Hamnett 2001: 160–1)

For many, in consequence, there is a pressing need for a 'policy turn' in social scientific enquiry (e.g. Markusen 1999, Martin 2001, Peck 1999). Beck (2000: 15) sums up the current situation as follows,

> We . . . are intellectual giants at picturing over and over again the endless chain of disasters and the impossibility of overcoming them. But we are

dwarfs when it comes to laying solutions on the table, or even spurring people on to conceive and test a way out of the horrors.

The rest of this book, in consequence, devotes itself to exploring the policy implications of our findings in Parts I and II. Until now, the overarching objective of the dominant third way discourse on poverty has been to achieve full-employment in the belief that employment is the best route out of poverty. As has been shown throughout this book, however, putting all of one's eggs into the basket of returning to the 'golden age' of full-employment is both illogical and unrealistic. Not only did this golden age never exist (since it was an age of full-employment for men only, not women) but the trend is ever further away from full-employment. Instead, a shift of economic activity from the market to the non-market sphere has been highlighted where the jobless also suffer relative inequalities so far as participation in informal work is concerned. In consequence, if poverty is to be tackled and a more inclusive society developed, it is our contention that it is necessary for policy to take on board the notion of enabling these populations to participate in the informal sphere to a greater extent than they presently find possible.

To pursue such an approach, an acceptance that the goal of full-employment is not feasible is first required and second, some alternative goal needs to be put in its place. This chapter thus reviews the rationales for shifting beyond the goal of full-employment and then outlines what is here considered to be an overarching goal that could replace it.

BEYOND THE GOAL OF FULL-EMPLOYMENT

Since the election of the Labour government in 1997 in the UK, there has been a strong reassertion of the desire to return to a full-employment society. Indeed, the overarching thrust of UK work and welfare policy is predicated upon such a desire. The belief is that the only way of achieving a more 'inclusive' society is to insert citizens into employment (Bennett and Walker 1998, Jordan 1998, Levitas 1998, McCormick and Oppenheim 1998, Powell 1999), and society has witnessed the ceaseless introduction of policy initiatives to try to realise this goal. Under the banner of 'making work pay', by which is meant making employment pay, there has been a rapid-fire salvo of new policy initiatives involving both 'carrots' and 'sticks' to entice people into formal jobs. The 'carrots' include: guaranteeing a minimum weekly wage through the tax credit system, introducing a 10 per cent starting rate of tax, 'modernising' National Insurance contributions, and introducing a minimum hourly wage. The chief 'stick', meanwhile, has been the introduction of 'welfare-to-work' policies such as the New Deals for young people, lone parents, the long-term unemployed and disabled (Bennett and Walker 1998, DSS 1998, Gregg *et al.* 1999, Hills 1998, HM Treasury 1997, 1998, Oppenheim 1998, Powell 1999). UK economic and social policy, therefore, similar to that in many other advanced economies (see Roche 2000, Van Berkel 2000), is dominated by a narrative that is based on achieving full-employment and that equates social inclusion with insertion into employment and social exclusion with unemployment.

Leaving aside the fact that all of these initiatives concentrate on increasing the supply of workers rather than job opportunities, the important point is that 'making work pay' has so far been interpreted only in a narrow manner to mean making it more viable for people to take formal employment. As we will argue below, another interpretation of 'making work pay' is to reward people for engaging in work in its widest sense. Indeed, for us, such a re-interpretation of 'making work pay' is absolutely essential if poverty is to be resolved and a more inclusive society developed.

As outlined in Part I, the size of the gap between actual employment rates and a full-employment scenario reveals a massive chasm that will need to be bridged if everybody is to be socially included through employment. This is particularly the case in deprived areas where the 'jobs gap' is even wider (Green and Owen 1998, Turok and Edge 1999). Indeed, this has not gone unrecognised by the governments of the advanced economies. Following the European Union's Lisbon Summit in 2000, all member states have now reasserted the goal of achieving 'full-employment' but have also redefined what they mean by such a goal. It is now accepted wisdom that 'full-employment' means achieving the same participation rate as competitor trading blocs, namely North America. A key issue for all member states, therefore, is what to do about the large segment of the population (about a quarter of all working-age people) who will remain jobless under this redefinition.

At present, there appears to be no coherent and explicit policy towards those who will remain excluded from employment. What, however, is to be done with them? And how are they to cope? For us, these questions can no longer be swept aside. Here, therefore, we sketch out a new goal to replace what we view as the crumbling edifices of a full-employment scenario.

TOWARDS FULL-ENGAGEMENT

Until now, as argued in Chapter 3, the dominant discourses on the future of work have all shared the view that paid employment is

> central to society, personal biographies and politics. If it is assumed, however, that the amount of paid employment is shrinking, then a change of paradigm or framework is required. The question then becomes which guiding idea, or ideas, will appear in place of the fixation on paid work. Or, to put it in another way, to what extent can alternative visions beyond the full-employment society already be glimpsed in people's living and working conditions and the projects they make for themselves?

(Beck 2000: 58)

This book has so far provided in its review of the coping practices of house-holds more than a glimpse of what might be this alternative vision. Running parallel to the formal sphere is a large (and growing) realm of informal work. Indeed, households heavily rely on this informal sphere to maintain and enhance their capabilities. For us, this realm provides a guide to the way forward. However, the informal sphere currently consolidates, rather than diminishes, the socio-spatial inequalities produced by formal employment. It thus becomes obvious that the informal as well as, or even instead of, the formal sphere must be harnessed if the needs and desires of deprived populations are to be met.

Here, in consequence, we assert that the goal of full-employment needs to be replaced by a new goal that recognises the importance of informal work in enhancing capabilities. For us, the replacement of 'full-employment' with the goal of 'full-engagement' enables this to be achieved. By 'full-engage-ment' is here meant providing sufficient work (both employment and informal economic activity) and income so as to give citizens the ability to satisfy both their basic material needs and creative potential.

There is recognition, however, that if a laissez-faire approach towards informal economic activity continues, then current socio-economic inequal-ities in households' abilities to perform such work will remain. As such, the achievement of a full-engagement society is predicated upon actively devel-oping the ability of people to help themselves and others so that they can choose how they wish to acquire goods and services to satisfy their needs and desires.

It is important to recognise that we are not advocating an approach that simply takes away the welfare 'safety net' and leaves people to fend alone. Instead, the approach here proposed differs in three respects to the tradi-tional 'laissez-faire' approach. First, it is based upon the logic of supple-menting and not substituting employment and state welfare provision. Second, it is based on the concept of optionality and choice, which contra-dicts the conservative appeal to duties and norms and third and finally, the promotion of informal work is envisaged more in terms of collective and interactive forms of working instead of in terms of isolation, compliance and competition.

In a full-engagement society, it is not an either/or choice between paid employment or informal work but a both/and approach. Given the evidence already presented, some may assert that this is already the case. This is correct in so far as people's everyday lives are concerned. The problem at present, however, is twofold. First, those least able to participate in the formal labour market are also those least able to draw upon the informal sphere for suste-nance and second, policy-makers have prioritised the development of the formal sphere over the informal sphere as the way forward.

The first step required in order to make the transition from a full-employ-ment to a full-engagement society, therefore, is to recognise and value work beyond employment. As Beck (2000: 58) puts it,

In the transition from the work [employment] society to the multi-activity society, a new answer is given to the question: what is work? The concept of an 'activity society' does, it is true, include a reference to a paid work, but only as one form of activity alongside others such as family work, parental work, work for oneself, voluntary work or political activity. This reminds us that people's everyday lives and work are stretched on the procrustean bed of *plural activities* – a self-evident fact that is usually obscured in the perspective of a society centred upon paid employment

Once informal modes of production are recognised and valued, the next step is to move towards giving it equal status to formal employment. At present, and as outlined in Chapter 3, this has not occurred. Although there is widespread recognition in the 'social' or welfare sphere that civil society is a third prong in the welfare equation that needs to be given equal status to the formal public and private spheres, the same reconceptualisation has not occurred in the so-called 'economic' sphere. Here, formal employment retains its status as the only form of work of any true worth and there is strong resistance to recognising and valuing informal modes of production as of equal value to formal modes of production. The result is that people are currently enabled to enter the world of employment but there is much less attention given to helping people to engage in informal modes of production.

If full-engagement is to be achieved, however, then there will be a need to re-balance the priority accorded to formal employment and to place as much emphasis on enabling people to engage in informal modes of production as is given to enticing them into the world of employment. If policy-making can achieve this, then as argued in Chapter 1, the desires of the population at large will be reflected. As Beck (2000: 106) once again puts it,

> more and more people are looking both for meaningful work and opportunities for commitment outside of work [employment]. If society can upgrade and reward such commitment and put it on a level with gainful employment, it can create both individual identity and social cohesion.

To do this, however, requires a rethinking of what constitutes economic progress. As the communitarian ecocentrists and radical European social democrats recognise so well, what were originally means to an end (e.g. employment, economic growth) have now become ends in themselves. What is required, therefore, is to recognise this and to revert to the original ends. As Giddens (2002: 39) asserts, 'Policies . . . should be focused upon developing people's capacities to pursue their well-being'. If this were adopted, the outcome would be that formal employment along with informal work would be assessed in relation to this objective. Today, this is far from the case. Indeed, the development of the informal sphere is usually evaluated purely in terms of its impacts on the development of the formal sphere, as if this were the ultimate goal. However, recognising that the worth of the informal sphere is not its impacts on the formal sphere but its impacts on household capabilities, then a new economic vision can start to be forged. Unless the current 'market-ism' that measures everything with reference to its impact on the formal sphere is overcome, there will continue to be a burgeoning desire to always uphold the values and priorities of the market. This will result in a continuation of the pressure being put on people doing informal work to take paid work.

A crucial issue here, moreover, is that advocating full-engagement must not be interpreted as a call to insert the 30 per cent of working-age people who are to remain jobless into a reconstituted 'informal' sphere whilst leaving the 70 per cent in employment unaffected. To do this will simply lead to a dual society in which the majority finds salvation through the formal sphere and the rest are confined to a second-class informal sphere to eke out an existence. By adopting full-engagement as the goal, the objective is to create a multi- or dual-activity society (e.g. Beck 2000, Gorz 1999). It is not to create a dual society.

Towards a dual-activity society, not a dual society

How, therefore, can a dual-activity society be realised without creating a dual society? To answer this, it is first necessary to recognise that a dual society (i.e. the polarisation of society into those who find their work and welfare through the formal sphere and those marginalised in the informal realm) is not the inevitable outcome of a dual-activity society (i.e. a society in which both formal and informal work are recognised and valued, and households are facilitated to engage in both forms of work in order to meet their needs and wants). The separation of the 30 per cent of the working-age population without a formal job into a second-class realm of work and welfare will only result if full-employment is retained as the goal and solutions are then sought for what is to be done with this surplus population.

Once full-employment is replaced with full-engagement, however, then the inevitability of a dual society is greatly reduced. The goal of full-engagement, to repeat, seeks to provide sufficient work (both employment and informal economic activity) and income so as to give citizens the ability to satisfy both their basic material needs and creative potential. It is thus not seeking to allocate some individuals and/or households to the informal sphere and others to the formal realm. Instead, it is pursuing a dual-activity society in which all households and/or individuals engage in both formal and informal work in order to meet their needs and desires.

Indeed, wherever a dual-activity society is advocated, this is the case. Beck (2000) highlights this in Germany amongst the various liberal, Green and communitarian groups pursuing a 'multi-activity' or 'dual-activity' society.

It is similarly the case in France where there is a long tradition of advocating such a future for work in a way that avoids the creation of a dual society (e.g. Aznar 1981, Delors 1979, Gorz 1999, Windebank, 1991). So too is such an argument prevalent in the UK (e.g. Boyle 1999, Douthwaite 1996, Jordan 1998, Mayo 1996) and the US (e.g. Cahn 2000).

Conscious of the fact that both old-style social democrats and neo-liberals might seize upon the idea of a dual-activity society so as to reduce formal welfare costs, dual-activity society advocates regard multi-activity as applying to the population as a whole rather than to solely the unemployed. As Beck (2000: 60) explicitly puts it,

> Only when every man and woman has one foot in paid employment, and perhaps the other in civil labour, will it be possible to avoid a situation where the 'third sector' . . . becomes a ghetto of the poor.

This is the central and essential requirement if a dual-activity society is to avoid the creation of a dual society. It is only by applying dual activity to the whole population rather than solely to the jobless that the creation a dual society can be avoided.

In practice, nevertheless, the problem is that it is precisely amongst the jobless that policy interventions are required in order to develop their capacities to engage in informal economic activity. Indeed, unless this is the focus of policy interventions, then the informal realm will continue to consolidate, rather than reduce, the socio-economic inequalities produced by the formal labour market. However, as soon as policy focuses upon the development of informal work amongst the unemployed, accusations will start to be raised that a dual society is being forged. There thus seems little that can be done to qualm such fears in the short run.

It is not only those who fear for the plight of the jobless, however, that have concerns about the creation of a dual-activity society. Another source of opposition to a dual-activity society is what we here call 'old-style feminists'. When seeking emancipation for women, much feminist literature retains a deeply ingrained belief that 'employment is the route to liberation' (see Gregory and Windebank 2000). Viewing informal work, whether in the form of housework or caring labour, as a 'burden' that unevenly falls on women's shoulders, the argument has been that insertion into employment is the royal road to emancipation.

As Hochschild (1989) has pointed out, however, this has been a 'stalled revolution'. Women, although entering the formal labour market in their masses, have remained responsible for the domestic workload. Indeed, when some women escape their domestic responsibilities it is only at the expense of other women, who are employed on low wages to free them. It is not men who have taken over this work. For many 'new-style' feminists, therefore, the route to women's liberation is being reconceptualised. It is not so much women needing to adopt work patterns more like men, but men adopting work patterns more akin to those conventionally associated with women. Only if this occurs will a dual society in which women remain responsible for the informal sphere be tackled (see Gregory and Windebank 2000). For these analysts, in consequence, the route to liberation for women is not through insertion into employment but through men adopting dual-activity lifestyles. Until this is accepted, one can expect a good deal of opposition from 'old-style feminists' who continue to cling to employment insertion as the route to women's liberation.

In this book, in consequence, our vision of a full-engagement society is one that adopts ways of working that are more akin to those which women have known for the past few decades than to those with which men are now familiar. They will not involve careers but rather combinations of part-time employment, casual contracts, unpaid work and volunteer activity for the public good.

The rest of this book primarily focuses upon policies to harness participation in informal modes of production so as to develop the capacities of people to meet their needs and desires. This is here adopted as our focus simply because there is so little literature on how this might be achieved. It is important to recognise, however, that the only reason for adopting this focus is that there is already so much literature on enhancing participation in formal employment. It is not meant to assert that we are not interested in this crucial question or that it is unimportant. Indeed, unless alterations in participation in the informal sphere are accompanied by radical changes in how people participate in the formal sphere, then the outcome will be a dual society. For example, leaving the distribution of employment concentrated amongst multiple-earner households with a large minority of households excluded from employment, and at the same time harnessing the ability of these latter households to engage in informal work, will inevitably create a dual society. For full-engagement to be achieved in a way that does not produce such a society, harnessing the informal sphere must march hand in hand with a redistribution of employment.

The key policy issue that needs to be dealt with in this regard is the more even distribution of formal employment as well as the income deriving from such employment. Throughout the advanced economies, a notable trend is that the total time that the population spends in employment is decreasing (see, for example, Gershuny 2000). This, moreover, is coupled with a process of social polarisation so far as the allocation of this time (and the income from it) is concerned. As the 'male breadwinner' model has declined, the advanced economies have increasingly polarised into multiple- and no-earner households. Unless the limited employment (and resulting income) is more evenly distributed, therefore, any policy to cultivate the informal sphere is likely to lead to a dual society. Several options are available in this regard. Possible policies to redistribute employment time include reducing the hours spent in employment so as to create a larger number of jobs, introducing a lifetime 'cap' on the number of hours spent in employment and sharing out some forms of employment (see, for example, Beck 2000, Gorz 1999, Windebank 1991). Although the focus here is primarily upon how to cultivate the informal sphere, these are issues that cannot be ignored. The informal side of the coin is here the focus, to repeat, merely because this is the realm that has so far received the least attention.

CONCLUSIONS

The aim of this chapter has been to argue that the goal of full-employment is redundant and that a new goal needs to be put in its place. For us, the notion of full-engagement is a fruitful avenue to pursue. This resonates with the desire to improve the capabilities of people to participate in the informal sphere and also enables the advanced economies to swim with the structural shifts currently taking place towards informalisation. Rather than put all one's eggs into the basket of full-employment as a policy solution, this chapter has thus argued that it is necessary to also develop informal practices if the coping capabilities of the poor are to be enhanced.

If the advanced economies manage to achieve full-employment, then the fact that this goal has been pursued will not be a problem. It will have played a positive role in facilitating greater community spirit and participation in civil society. If full-employment is not achieved, however, then this approach will have created alternative means of meeting needs and desires. As stated, nevertheless, doing this in isolation from changes in the formal sphere is not possible. To pursue the development of such additional poverty alleviation policies without changing the allocation of work in the formal sphere will act merely to create a dual society. Given the lack of attention so far devoted to harnessing participation in informal work compared with the amount of attention paid to inserting people into employment, the next two chapters focus upon how this might be achieved. To do this, we consider first, bottom-up initiatives and second, top-down initiatives that can be used to cultivate participation in this work.

8 The new mutualism

A fourth sector approach

Can bottom-up or grass roots initiatives be used to alleviate poverty? If so, what type of initiatives need to be developed and for what purpose? In the employment-centred discourses of conventional third way thinking, bottom-up initiatives are seen to play a central role (e.g. Blair 1998, Giddens 1998, Leadbeater 1999). Focusing upon more organised third sector initiatives, the argument is that these can provide additional springboards into employment for the unemployed so as to fill the gaps left by the private and public sectors and/or new forms of voluntarism to meet welfare needs.

Throughout this book, however, it has been argued that if poverty is to be alleviated, there is a need to enhance the capacities of people to engage in informal economic activity as an 'economic' alternative to formal employment (see, for example, Chapter 3). Here, in consequence, a rather different approach is adopted than in conventional third way thought with regard to developing bottom-up or grassroots initiatives both in terms of the types of association focused upon and the rationales for developing them. Rather than develop the third sector (organised community-based associations) as a springboard into employment, the approach towards cultivating bottom-up initiatives here develops what we call the 'fourth sector' (micro-level informal economic activity such as one-to-one reciprocity) in order to develop alternative means of livelihood beyond employment.

To commence this chapter, therefore, we outline the contrasting interpretations of the role of bottom-up initiatives in paving a third way. This will provide an understanding of how a 'new mutualism' based on harnessing the fourth sector in order to achieve fuller-engagement radically differs from the conventional third way approach that seeks to develop the third sector in order to move towards full-employment. Following this, the various initiatives that could be used to develop the fourth sector in order to provide alternative means of livelihood beyond employment are considered. Each initiative reviewed here is evaluated due to its strong resonance with the contemporary attitudes towards participation in informal work identified in Chapter 5. They each focus upon developing the capacity of people to engage in informal economic activity by using some form of tally or accounting system so that people can keep tabs on how much they are giving and receiving. As revealed earlier, this is precisely what people desire when engaging in acts of exchange with others.

These initiatives are local exchange and trading schemes (LETS), time banks, employee mutuals and mutual aid contracts. Our argument is that if these initiatives are developed, then the capacity of people to engage in informal economic activity as an alternative to employment can be enhanced. However, throughout this chapter, it is revealed that if these initiatives are evaluated in conventional third way terms as springboards into employment, then they are relatively ineffective. It is only when they are evaluated as instruments for harnessing work beyond employment that both their effectiveness and their central role in alleviating poverty and creating a more inclusive society becomes apparent.

INTERPRETATIONS OF THE ROLE OF BOTTOM-UP INITIATIVES IN PAVING A THIRD WAY

The centrality being accorded to bottom-up initiatives in paving a third way has been seen as necessary for a variety of reasons (see, for example, Beck 2000, Giddens 1998, 2000, Hargreaves 1998, Jordan 1998, Powell 1999). Whether analysts adopt full-employment or full-engagement as their goal, nevertheless, all agree that one of the most significant stimuli for rethinking the role of bottom-up initiatives is the employment problem facing the advanced economies. Such bottom-up initiatives are seen as providing a potential solution. Situated beyond the private and public sectors of the economy, these initiatives comprise private formal associations that pursue economics-orientated collective self-help based on not-for-profit and co-operative principles (e.g. Chanan 1999, Lorendahl 1997, Pestoff 1996, Westerdahl and Westlund 1998). Often, informal networks of the family, kin, neighbourhood and community, or what might be called 'fourth sector' associations, have been excluded from consideration under the banner of bottom-up or social economy initiatives (e.g. Amin *et al.* 2002). Here, however, and given our findings on the centrality of such associations in coping capabilities and practices of lower-income populations, they are included in our discussions. After all, such micro-level one-to-one exchanges are just as much a form of economics-orientated collective self-help as the formal community-based groups that are the usual focus of enquiry.

How, therefore, can bottom-up initiatives resolve the 'employment problem'? Until now, two distinct answers have been given to this question. Here, each will be outlined in turn in terms of the role that the bottom-up initiatives are seen to play and the criteria used to evaluate the effectiveness of such initiatives.

The 'old mutualism': the third sector as a job creator and springboard into employment

This first approach focuses upon the third sector and views it as a means filling the jobs gap left by the public and private sectors (Archibugi 2000, Community Development Foundation 1995, ECOTEC 1998, European Commission 1996b, 1997, 1998b, Fordham 1995, OECD 1996). Here, therefore, the emphasis is upon whether the third sector can generate jobs and improve employability so as to facilitate a return to full-employment.

In an age of de-coupling of productivity increases from employment growth, the perception is that the private sector can no longer be relied upon to create sufficient jobs. Neither, moreover, can the post-war corporatist welfare state model be expected to spend its way out of economic problems. In this context, the third sector is seen as a potential solution. It is thus bolted on to conventional job creation programmes and policies. Indeed, this approach has steadily gained momentum not only throughout Europe but also North America (European Commission 1996b, 1998b, Mayer and Katz 1985). As evidence of its prevalence, one has only to note that the European Commission's major mechanism to stimulate the third sector is entitled the 'Third System and Employment' (see ECOTEC 1998, Haughton 1998, Westerdahl and Westlund 1998).

The development of the third sector is thus seen to complement the current range of carrots and sticks used under the umbrella term of 'making work pay' that seeks to increase the numbers available for employment. This includes tax credits, the minimum wage, national insurance and tax modernisation as well as the 'welfare to work' programmes (e.g. Bennett and Walker 1998, DSS 1998, Gregg *et al.* 1999, Hills 1998, HM Treasury 1997, 1998, Oppenheim 1998, Powell 1999). The development of the third sector complements these policies by encouraging people to take jobs in two ways. On the one hand, it provides additional job opportunities to those created by the public and private sectors. On the other hand, it improves the employability of those excluded from the public and private sectors by maintaining and enhancing their work-related skills. In this sense, the third sector is an essential supplement to the current 'making work pay' policy agenda. A job requires not only a person to be available but also a job opportunity and a suitably qualified person. The role of the third sector is to both provide these additional job opportunities and improve employability by helping people to maintain and acquire skills, and develop self-confidence and self-esteem.

To evaluate the effectiveness of the third sector, therefore, whether it creates jobs and improves employability is assessed. Evaluation criteria include the number of formal jobs created by an initiative, its ability to facilitate skill acquisition and maintenance, whether it provides a test bed for new potential formal businesses, its ability to develop self-esteem, and its ability to maintain the employment ethic.

The 'new mutualism': fourth sector initiatives as an instrument for facilitating alternative means of livelihood

Others, however, ascribe a rather different role to bottom-up initiatives. Rather than focus upon third sector initiatives as part of a strategy to achieve 'full-employment', this perspective instead seeks to cultivate the 'fourth sector' in order to facilitate 'full-engagement' (see Jordan *et al.* 2000, Mayo 1996, Williams and Windebank 1999b). As such, the role of bottom-up initiatives is not to create jobs and improve employability but to facilitate alternative means of livelihood in the form of informal economic activity. At its heart is an understanding of the need to reduce the perceived importance attached to conventional employment and recognise people's broader social contributions by valuing the vast and growing amount of informal work that takes place in society (cf. Levitas 1998, Lister 1997).

However, it is also understood that unless the laissez-faire approach towards informal work is transcended and proactive policies are developed to cultivate such activity, the exploitation and socio-economic inequalities inherent in such work will continue to prevail (Williams and Windebank 1999b, 2000a). In order to improve what Sen (1998) calls the 'capabilities' of people to help themselves, bottom-up initiatives are thus sought that can facilitate participation in informal work and stem the degradation of the social fabric in terms of the capability for reciprocal exchange (e.g. Chanan 1999, Macfarlane 1996, OECD 1996).

The intention is to develop bottom-up initiatives that can tackle the barriers to participation in informal economic activity identified in Chapter 6. These include the fact that people lack, first, the money to acquire the goods and resources necessary to participate in reciprocal exchange (economic capital). Second, they know few people well enough to either ask or be asked to do something (social network capital). Third, they lack the appropriate skills, confidence and/or physical ability to engage in self-help (human capital) and fourth and finally, they fear being reported to the tax and/or benefit authorities if they engage in such work (an institutional barrier). To evaluate the potential of bottom-up initiatives in this approach, therefore, a different set of criteria is used. Initiatives are evaluated in terms of their ability to tackle the barriers to participation in informal work.

Below, a range of bottom-up initiatives are documented that attempt to both provide alternative means of livelihood and do so in a manner that resonates with the contemporary attitudes towards reciprocal exchange identified in Chapter 5. These are local exchange and trading schemes (LETS), time banks, employee mutuals and mutual aid contracts. In each case, we highlight how although these initiatives may not be efficient and effective if full-employment is the goal, this is not the case if the goal is fuller-engagement.

LOCAL EXCHANGE AND TRADING SCHEMES (LETS)

In recent years, a host of UK government policy documents have tentatively suggested that LETS might be a means of tackling poverty and unemployment (e.g. DETR 1998, DfEE 1999, Home Office 1999, Social Exclusion Unit 2000). In the new spirit of 'evidence-based policy-making', however, UK government departments have held back from advocating these initiatives until evidence has been gathered and reported on their effectiveness in tackling poverty and social exclusion. As Boyle (2002) asserts, discussions with senior civil servants about the future role of LETS as a means of tackling poverty and social exclusion have been continuously met with the response that evidence was awaited on their contribution to achieving this objective. This evidence gathered under an ESRC grant on LETS as a means of tackling social exclusion was eventually reported in late 2001. Here, we review the results of this study (Williams *et al.* 2001).

Until this study, the only evidence available on their impacts was in the form of evaluations of individual LETS (e.g. Barnes *et al.* 1996, Lee 1996, North 1996, 1998, 1999, Pacione 1997, Williams 1996a, 1996b, 1996c). The value of the recent study is that it is a comprehensive national evaluation of LETS. To report the results, first, the effectiveness of LETS in creating jobs and improving employability, and second, their effectiveness as instruments for facilitating alternative means of livelihood, will be examined.

LETS arise when a group of people form an association and create a local unit of exchange. Members then list their offers of, and requests for, goods and services in a directory that they exchange priced in a local unit of currency. Individuals decide what they want to trade, who they want to trade with, and how much trade they wish to engage in. The price is agreed between the buyer and seller. The association keeps a record of the transactions by means of a system of cheques written in the local LETS units. Every time a transaction is made, these cheques are sent to the treasurer who works in a similar manner to a bank sending out regular statements of account to the members. No actual cash is issued since all transactions are by cheque and no interest is charged or paid. The level of LETS units exchanged is thus entirely dependent upon the extent of trading undertaken. Neither does one need to earn money before one can spend it. Credit is freely available and interest-free. As such, LETS are very much third sector initiatives. They are private formal associations for pursuing economica-orientated collective self-help based on not-for-profit and co-operative principles.

To evaluate the effectiveness of LETS both at helping members into formal jobs and as tools for facilitating alternative means of livelihood, three methods were employed in this study. First, a postal survey of all LETS co-ordinators was undertaken in 1999. Of the 303 LETS identified and surveyed, 113 responded (37 per cent). Second, a membership survey was conducted with some 2,515 postal questionnaires being sent out and 810 (34 per cent) returned. Third and finally, in-depth action-orientated ethnographic research was conducted on two LETS in very different locations: the semi-rural area of Stroud and the deprived urban area of Brixton in London. Here, we report the results.

The social bases of LETS

Starting with the overall magnitude of LETS, the co-ordinators survey identified that the LETS responding had an average of 71.5 members and a mean turnover equivalent to £4,664. If these LETS are taken as representative, then the total UK LETS membership is 21,664 with a turnover equivalent to some £1.4 million. In terms of total exchange-value, therefore, LETS are relatively insignificant. However, when measured in terms of their use-value, as we shall show below, this is not the case.

Who, therefore, joins LETS? Of the 810 members responding to the survey, LETS members are predominantly aged 30–49, women, relatively low-income groups and those who are either not employed or are self-employed. Indeed, if non-employment and low household incomes are taken as surrogate indicators of social exclusion, then the membership is heavily skewed towards the socially excluded. Some 62 per cent of members were not employed and 66 per cent of all members lived in households with a gross income of less than £20,000. On the whole, however, they constitute a specific segment of the unemployed and those living on low incomes. Due to the origins of LETS in the 'alternative' and green movements, they are mostly those who are disenchanted with employment and seeking to downshift.

Why, therefore, do these people join LETS? Some 25.2 per cent do so for ideological purposes. LETS for them are 'expressive communities': acts of political protest and resistance to the 'mainstream' where ideals can be put into practice (cf. Hetherington 1998, Urry 2000). Of the remainder, just 2.5 per cent join explicitly to improve their employability. The remaining 72.3 per cent join first for 'social' reasons (22.9 per cent), such as to meet people, rebuild a sense of community or help others, and second for 'economic' reasons (49.4 per cent), such as to help them overcome their lack of money or exchange goods and services. 'Social'/community-building reasons tend to be cited by the employed and relatively affluent and economic reasons by the relatively poor and non-employed.

Given this concentration of low-income households and non-employed people in the membership, albeit from a particular segment, and the fact that many join for economic reasons, the effectiveness of LETS is now evaluated first, as job creators and springboards into employment and second, as instruments for facilitating alternative means of livelihood.

Evaluating LETS as a job creator and springboard into employment

To evaluate LETS in this regard, one must examine the number of formal jobs created, its ability to facilitate skills acquisition and maintenance, whether it provides a test bed for new potential formal businesses, its ability to develop self-esteem and its ability to maintain the employment ethic.

Government has been tentatively supportive of LETS because it was seen as a potential means of inserting people into employment (e.g. DfEE 1999, Social Exclusion Unit 2000). However, this third sector initiative is not a direct job creator. The number of direct jobs created amounts to a dozen or so since volunteers mostly run them. Nor do the LETS *directly* improve employability. Just 5 per cent of respondents said LETS had directly helped them gain formal employment. Working in the LETS office administering the scheme had enabled valuable administrative skills to be acquired, which had been used to successfully apply for formal jobs. Their ability in this regard is thus limited: only a small number of people can at any one time play a prominent role in administering the scheme.

However, they are effective at *indirectly* improving employability. Some 27 per cent of all respondents asserted that the LETS had boosted their self-confidence (33.3 per cent of the registered unemployed) and 15 per cent per cent that new skills had been acquired (24.3 per cent of the registered unemployed), mostly related to computing, administration and interpersonal skills.

Furthermore, it is not just employability that is improved on LETS. Some 10.7 per cent of members asserted that LETS had provided them with a useful seedbed for developing their self-employed business ventures. It had enabled them to develop their client base (cited by 41.2 per cent of those who were self-employed), ease the cash flow of their business (cited by 28.6 per cent) and provided a test bed for their products and services, cited by nearly all who defined themselves as self-employed. LETS, in consequence, although not a job creator, does provide a useful springboard into employment and self-employment for a small but significant proportion of members.

Within the logic of this approach towards the third sector, therefore, several possible policy responses arise. To further enhance the capacity of this third sector initiative as a generator of employment, first, the 'voluntary and community sector' of the New Deal programme could be used to fund LETS office workers for their administrative work. This would provide workers with a proven means of entering the formal labour market as employees and, at the same time, enable the more efficient running of the LETS (since it would not be so reliant on volunteers for its day-to-day administration). Second, many currently operating as self-employed in LETS could be encouraged to enter the 'self-employment' option in New Deal and their trading on LETS could be recognised as part of their attempt to become self-employed. If these steps were taken, then the effectiveness of LETS in paving a conventional third way could be improved. Interestingly, however, it appears unlikely that either of these initiatives will be pursued.

'Off the record' feedback from senior civil servants to the results of this survey has suggested that because LETS are relatively ineffective at creating jobs and improving employability, they are unlikely to receive support from government. This not only reveals the 'employment-centred' discourses of current third way thinking but also the way in which such bottom-up initiatives are only measured at present in terms of their effectiveness in this regard. As we have argued throughout this book, however, evaluating such initiatives only in terms of their effectiveness in relation to spiriting people into the formal labour market is short-sighted.

Evaluating LETS as an instrument for facilitating alternative means of livelihood

In the 'new mutualism', LETS are evaluated not in terms of their effectiveness as springboards into employment but in terms of whether they facilitate alternative means of livelihood beyond the formal sphere. To do this, these bottom-up initiatives are analysed in terms of the extent to which they counter at least four key barriers to participation in work beyond employment.

Starting with the extent to which LETS tackle the barrier of economic capital, some 40 per cent of members assert that LETS provided them with access to interest-free credit (but 62.1 per cent of the registered unemployed and 51 per cent of low-income households). LETS, therefore, provide people with access to money. For two-thirds (64.5 per cent) of the registered unemployed, this had helped them cope with unemployment, with some 3.1 per cent of their total income coming from their LETS activity.

LETS also enable the barrier of social network capital to be tackled. Some 76.2 per cent of respondents asserted that the LETS had helped them to develop a network of people upon whom they could call for help whilst 55.6 per cent asserted that it had helped them develop a wider network of friends and 31.2 per cent deeper friendships. LETS, therefore, develop 'bridges' (i.e. bringing people together who did not before know each other) more than 'bonds' (i.e. bringing people who already know each other closer together). Given that most members lacked kinship networks in the localities they inhabited and that kinship networks are the principal source of mutual aid in contemporary society (Williams and Windebank 1999b), LETS thus provide those without such a local network with a substitute. Some 95.3 per cent of LETS members, that is, had no grandparents living in the area, 79.5 per cent no parents, 84.3 per cent no brothers or sisters, 58.2 per cent no children, 92.6 per cent no uncles or aunts and 90.8 per cent no cousins.

Besides tackling the barriers of economic and social network capital, there is also evidence that LETS tackle the barrier of human capital that can constrain participation in reciprocity. As discussed above, LETS provide an opportunity for people both to maintain and develop their skills as well as to rebuild their self-confidence and self-esteem by engaging in meaningful and productive activity that is valued and recognised by others who display a willingness to pay for such endeavour.

Finally, there is an institutional barrier to engaging in alternative means of livelihood. Many who are unemployed are fearful of being reported to the authorities, even if they engage in unpaid mutual aid. This is not currently being overcome by LETS. Although only 13 per cent of members feel worried about tax liabilities, 65 per cent of registered-unemployed members are concerned about their situation. Moreover, all of the registered unemployed interviewed not currently involved in LETS were put off joining by worries about its impacts on their benefit payments. Ironically, therefore, those who would most benefit from LETS are discouraged from joining and trading due to the uncertainty over their legal position vis-à-vis the benefits disregard. The current 'laissez-faire' approach of government, in consequence, is insufficient to appease both members and non-members who are registered unemployed.

For us, therefore, key idea 4 of the Social Exclusion Unit in its *National Strategy for Neighbourhood Renewal: a framework for consultation* should be supported. This proposes a pilot study to give 'people new freedom to earn a little casual income or participate in a Local Exchange and Trading Scheme (LETS) without affecting their benefit entitlement' (Social Exclusion Unit 2000). Even though LETS may not be highly effective as springboards into employment, they are relatively more effective in developing alternative means of livelihood in that they tackle the barriers to participation in such activity. To enable LETS to become more effective in this regard, however, several changes are required.

Barriers to the development of LETS as alternative means of livelihood

Conventional third way thinking has much to learn from LETS. Rather than rely on formal employment as the sole route out of poverty, the members of these pioneering initiatives are adopting a 'work ethic' rather than an 'employment ethic' in order to alleviate their situation. They are using informal exchange as an alternative means of livelihood. Government needs to follow suit. By harnessing bottom-up initiatives as a tool for encouraging informal exchange (rather than job creation) the full potential of such initiatives could start to be realised. By recognising and valuing work beyond employment, government would be taking the first step to recognising that the basic material needs and creative desires of its citizens could be met just as well by pursuing 'full-engagement' rather than full-employment.

Once this is recognised, then several additional barriers will need to be overcome for a wider proportion of the population to enjoy the benefits of these initiatives. To see this, we here address five questions that confront individuals currently wishing to join LETS:

- does a LETS exist in my locality?
- do I know about it?
- do I feel that it is something for me?
- do I feel that I have something to contribute?, and
- do I fear the backlash of the tax and benefit authorities if I engage?

Here, we consider the present answers to each question to show the work required if LETS are to become more effective tools for tackling social exclusion.

Does a LETS exist in my locality?

Many localities do not have LETS in their area. The co-ordinators survey reveals that the average LETS covers 126 square kilometres. Those 303 LETS currently existing in the UK thus cover only 38,178 square kilometres, or 15.6 per cent of the UK. Hence, for people in most areas, the principal barrier to their participation is that LETS do not exist.

Do I know about it?

During research in Stroud, a relatively small tight-knit town that possesses one of the longest-standing and largest LETS in the UK, 108 people were interviewed who were not LETS members. Of these, 51 per cent had never heard of LETS. If over half of the people in a town such as Stroud, where there is an established scheme (with a high-street presence until recently) and a relatively tight-knit community, have not heard of the LETS, then this figure must be considerably higher in other localities. Indeed, and as evidence that this is the case, a repetition of this survey in Brixton revealed that some 90 per cent of people interviewed had never heard of LETS. There is a good deal of work required, therefore, to extend knowledge of LETS to a wider range of people.

Do I feel that it is something for me?

If one exists and people know about it, the next question is whether it is something for them. In our survey, two-thirds (67 per cent) of non-members thought that it was not. First, this is because they were either 'money rich but time poor' or had extensive kinship networks that substituted for LETS. Second, however, it is because they either fear having their social benefits curtailed (dealt with below) or perceive LETS as something for people other than them. Indeed, given that 61.3 per cent of members hold graduate or above qualifications and 48 per cent support the Green Party, it appears that these inclusive mechanisms for some are exclusionary for others (see also Lee 1996, Williams 1996c).

The solution, however, is relatively simple. This membership profile has arisen because of the way that LETS advertise themselves. The principal first instance method used by LETS when seeking new members is 'word of mouth' (employed as the principal marketing device by 55 per cent of LETS), followed by public posters/flyers (19 per cent), newspaper coverage (9 per cent), public stalls (6 per cent), targeted mail shots (4 per cent), talks to local groups (3 per cent), advertising in shop windows (2 per cent) and leaflets to each household (1 per cent). In nearly all cases, moreover, LETS aim their publicity in the first instance at groups likely to be interested, such as environmental organisations, so as to pursue the 'line of least resistance'. The result is a skewed membership profile with many 'greens' and 'alternative lifestylers' joining but few people from other social groups. The result is intransigence in membership profiles since many then perceive LETS as something for others rather than them. Conscious recognition of how these advertising and recruitment practices lead to exclusion of other social groups is the first step in resolving this problem. The next is to adopt strategies when seeking new members that resonate with the groups sought (e.g. choosing appropriate locations for trading and social events, designing targeted promotional material).

Do I feel that I have something to contribute?

If one exists, people know about it and view it as something for them, they might still not wish to join because they do not feel that they can contribute anything. Indeed, this came across strongly in our interviews with non-members. For instance, the elderly and disabled felt that they could do little as did the unemployed. At the advertising stage, therefore, concrete examples are needed of not only what people can get on LETS but also what they can contribute. For example, cameos of how individuals belonging to the group targeted currently trade would be useful. Once joined, moreover, proactive policies are required to enable people to recognise their skills as well as acquire and develop new ones.

Do I fear a backlash from the tax and benefit authorities if I join and trade?

Finally, if LETS are to be effective vehicles for tackling social exclusion, then people must not fear recourse from the tax and benefit authorities. Currently, this is most widely felt amongst the unemployed, yet this is the group that perhaps more than any other could benefit from LETS. The changes required, however, depend upon the perspective towards social exclusion adopted. We have dealt above with how LETS could become a more efficient springboard into employment. To make LETS more efficient facilitators of alternative means of livelihood, however, requires the implementation of measures such as the recently failed LETS and Social Security Bill to enable the unemployed to participate without fear of recourse. It also requires greater resources for LETS development. Currently, just 49 per cent of LETS surveyed receive assistance, of which 76 per cent comes from local government. If LETS are to enable alternative means of livelihood, this funding will not only have to become more widespread and continuous but need to move beyond employment-related evaluation criteria towards alternative measures, such as the number of additional trades conducted either overall or else by specific social groups. At present, however, this seems unlikely. So long as an 'employment ethic' rather than a 'work ethic' persists, it seems evident that LETS will fail to receive much political support as a bottom-up initiative.

TIME BANKS

Another initiative that has the potential to tackle the barriers to participation in informal economic activity in a way that resonates with contemporary attitudes towards reciprocal exchange is the time banks scheme. Here, participants are paid one 'hour' for each hour that they work, which they can at any time 'cash in' by requesting an hour's work in return from the system (see Boyle 1999, Cahn 1994, 2000, Cahn and Rowe 1992).

This 'time currency' approach commenced in the mid-1980s in the US, when Edgar Cahn developed the idea of rewarding people for every hour of their community service in Time Dollars. The idea was to create a currency to record, store and reward transactions where neighbours help neighbours. People earn time currency by helping others (e.g. by providing child- or eldercare, transportation, cooking, home improvement). They then spend time currency to get help themselves or for their families, or to join a club that gives them discounts on food or health care.

As such, time currency allows those aspects of people's lives for which the market economy assigns no value to become redefined as valued contributions, and it gives society a way to recompense activities that the market economy does not. This time currency thus empowers people to convert personal time into purchasing power, so as to enable them to stretch their limited cash further. It also reinforces reciprocity and trust, and rewards civic engagement and acts of decency in a way that generates social capital, one hour at a time. The result of valuing such work is that it can help harness underutilised human resources, give value and recognition to activities that are currently unvalued and unrecognised, and generate social capital in communities by rebuilding the non-market economy of family, neighbourhood and community.

What is so useful about time banks is that, at present, governments can only resort to calls for greater 'civic engagement' and 'community involvement'. They have no way of rewarding such activity on the part of participants. Time banks, however, provide a means by which people can be recompensed for such activity. They reward reciprocity and convert that contribution into a form of currency that can be used to acquire goods and services that one needs or desires.

This volunteer service credit approach has received less attention in the UK compared with LETS, although this is now changing. The New Economics Foundation has started to co-ordinate the growing interest in 'time money' and pilot schemes are being developed in Camden, Lewisham, Gloucestershire (Fair Shares in Newent), Newcastle, Peckham and Watford, some with the help of National Lottery funding. Until now, however, there has been no evaluation of time banks either as springboards into employment or as vehicles for facilitating alternative means of livelihood. Here, therefore, we attempt to do so.

Time banks as springboards into employment

By 1998, over two hundred time banks and service credit programmes were operating in 30 states in the US and these schemes frequently have thousands of members. Crucially, and in contrast to LETS, these are professionally managed projects, requiring around $50,000 per annum to run a central office so as to match the needs of members with the volunteers available (Boyle 1999). To understand both the size and nature of time dollar schemes in the US, we here provide some brief sketches of just a few of the schemes in existence:

- In New York, a time dollar scheme called Elderplan focuses on the provision of health care for senior citizens. In 1999, 97,623 time dollars were earned serving 4,316 members through 41,985 care-giving episodes.
- In Chicago, a time dollar scheme centred on the school system enables Chicago to be able to boast that it possesses the nation's largest after-school cross-age peer tutoring programme. In its fourth year of operation in 2000, it had spread to 25 schools. Older students tutor younger ones and earn time dollars for doing so. In the academic year 1999–2000, 1,500 students participated in this scheme.
- In Baltimore's Hope VI project, the rent for accommodation includes the payment of 8 time dollars per month. This has encouraged 150 households to provide help to each other, to the local school and to their community. Families use additional time dollars earned to buy their bus pass, get discounts at shops, purchase furniture, clothing and membership at the Boys and Girls Club.
- The St Louis Grace Hill Neighbourhood Services programme uses time dollars and in the first ten months of 1999, there were 12,378 exchanges involving 42,519 time dollars.

Based on such examples of their potential cited in Cahn (2000), considerable attention has been given to these schemes in recent years from other nations. One of the first countries to try to replicate these schemes was the UK where, in the past few years, time currency schemes have started to be developed.

Unlike with LETS, the Benefits Agency in the UK has also decided that the 'credits' earned on time banks should not be treated as earnings for income-related benefit purposes. Participation in time banks is not seen as remunerative work and entitlement to benefits is not affected. Similarly, these credits are ignored for income tax purposes. Although such tax and benefit rulings have been made in the UK and support has been given by the Home Office's Active Community Unit for the development of the scheme, in mid-2001 there were still only 15 time banks operating and 21 under development. These had 400 participants who have earned 9,760 hours. As such, time banks remain relatively small compared with LETS. However, there is little doubt that they are rapidly expanding. Indeed, the objective of the New Economics Foundation who are co-ordinating their development in the UK is to have 120 operating by 2003.

Given this desire to develop them, a key question is whether time banks operate as a springboard into employment? So far, there has been little formal evaluation in either the US or the UK of the ability or potential of this initiative. Such research is badly required. Similar to LETS, however, the likelihood appears to be that they will create only a few dozen jobs directly by employing people to administer the time bank. Their greatest contribution as a springboard into employment will be in enabling skills maintenance, skills acquisition, the maintenance of the work ethic and building self-confidence.

Focus group research conducted on the first time bank to be set up in the UK, Fair Shares in Gloucestershire, however, suggests that participants tend to view time banks much more as community-building vehicles than as springboards into employment (see Williams *et al.* 2001). As such, their potential as springboards into employment will be more an offshoot of their main function of facilitating alternative means of livelihood than an explicit objective behind their creation. One exception is those time banks currently being created in cities such as Leicester that call themselves 'skill swaps'. These are in part explicitly being created to enable participants to acquire new skills and to act as a test bed for participants' self-employed business ventures. In these skill swap schemes, people will earn 'learning credits' for the skills they share with other people and use the credits they earn to buy their next steps up the training ladder. These will need to be evaluated as the pilot projects come to fruition in terms of their potential as springboards into employment and self-employment.

Time banks as alternative means of livelihood

Do time banks resonate with contemporary attitudes in the UK concerning the giving and receiving of help and thus facilitate the development of alternative means of livelihood? To investigate this, in 2000, a case study was undertaken on the first and largest time bank that is currently in existence in the UK: Fair Shares in Gloucestershire. Here, we report the results.

A case study of Fair Shares, Gloucestershire

As part of the ESRC project on LETS (see Williams *et al.* 2001), three focus groups were conducted with participants in Fair Shares time bank in Gloucestershire, which had by then been in operation for three years and had several hundred members.

An important finding from these focus groups is that people who join time banks not only possess similar preferences to those identified in Chapter 5 in relation to giving and receiving help but also join this scheme precisely because it adheres to these preferences. Many focus group participants, for example, were adamant that they did not wish to receive charity and neither would they offer one-way giving to others. Indeed, the fact that time banks avoided such one-way giving and receiving was seen as a major advantage of such schemes. As one focus group participant put it,

> The thing with Fair Shares is that . . . you can ask for a service and it will be done but you feel that because you are giving something back. You are not having something from like a charity, because a lot of people are like, 'Oh I don't want charity, I don't want charity, I don't want help', but through Fair Shares you get what you need but you can give back, so it sort of balances out really.

It is not only that time banks avoid the notion of charity, however, that makes them popular. It is also the way in which helping others is reciprocal and remunerated. This was well portrayed by three participants during one of the focus groups:

Participant A: If someone gets something for nothing, they either resent the person that's given it, or they resent – they get up tight about accepting. But if they feel they can give something back at a future date, then they are more likely to ask and that's what Fair Shares is about.

Participant B: But you don't feel obligated to anyone when you ask for help, you don't feel obligated because you know that they're earning credits that you are also giving them something that they can use. I mean that's the idea isn't it?

Participant C: And it also means that someone who is very tight for cash and that's most of us anyway doesn't have to put their hand in their pocket to get that help. They can call on somebody to help them without thinking 'Oh how much is this going to cost me?'

Indeed, for nearly all participants, this latter ability of time banks to enable help to be given and received without having to pay with national currency that is in short supply was a principal benefit. Take, for example, this conversation from one of the focus groups when discussing this advantage of having a new form of money beyond the national currency with which to engage in transactions.

Participant A: At least you're not paying actual cash for it which a lot of people can't afford.
Participant B: I mean it puts it on a completely different footing really.
Participant C: It does completely, yeah.
Participant D: Which is a good thing,
Participant E: It's doing favours for each other.
Interviewer: So does it make it easier to ask?
Participant E: Yes.
Participant C: Yes, yes.

For many participants, the time bank was thus a way of getting jobs done that, although seen by them as necessary, would not have been completed without the existence of the time bank. Take, for example, the following discussion:

Participant A: It's a way of finding people that can do the jobs that you wouldn't otherwise be able to do. I cannot afford to pay market rates and don't know who else to go to but by joining this scheme, the Office does know people and will put you in touch. It takes a worry off my mind.
Participant B: Yep, the fan broke on my oven. They found this electrician who was a member who had a fault-finding machine and he came along and sorted that out because one of the switches wasn't working . . . It's quite a good job done really. I had a good exchange there really because he saw my ladders and asked if he could borrow them. It didn't involve me in anything really . . . just lending the ladders. He came and fetched them on his roof-rack and borrowed them for a week.

It is not solely the fact that necessary work can be completed in a way that resonates with contemporary UK preferences for remunerating those who help you that was the advantage of time banks. For some, it was also the fact that everybody's time is treated equally. As one participant summarised,

I agree with everything people have so far said but for me in addition to everything else, one of the things that I found attractive about the scheme was that it equates people's time and I think that's a really interesting thing that an hour of my time is worth an hour of anybody else's time and I think that's quite a revolutionary idea and I think it could go a hell of a long way and it's quite refreshing for me . . . It is tied up with self-worth isn't it and because you charge forty pounds an hour and somebody else charges two pounds an hour or the minimum wage or whatever it is, therefore the person who gets forty pounds an hour is better or – that is somehow a kind of societal pressure and so it's quite an important scheme for re-addressing that idea of self worth. So that's the reason why I think it's a great scheme . . . it is a political movement in the best possible sense.

In sum, these focus groups with participants in the national pilot project for UK time banks reveal that the scheme resonates well with contemporary UK values towards helping others discussed in Chapter 5. Time banks avoid one-way giving and receiving, and provide a means of remunerating those who give and receive help. For most participants, therefore, time banks provide not only a vehicle that resonates with their normative values about giving and receiving help but also an instrumental means of getting necessary work completed. For some participants, moreover, the fact that they forge alternative constructions of value and different social relations of work and exchange beyond the profit motive and market relations are an added advantage.

It was earlier shown that, beyond kin, people do not want to receive charity and are wary of giving and receiving help if they are going to feel that they owe an individual, or s/he owes them. It has been here revealed that time banks reflect these preferences. Time currencies provide some form of remuneration for helping others and treat such acts as reciprocal in nature. Indeed, examining the attitudes of participants in Fair Shares has shown that people join precisely because it resonates so well with their preferences in relation to giving and receiving help. It can be thus concluded that time banks are transferable to the UK situation. They represent a potentially valuable tool for enabling people to help each other out in a way that conforms to contemporary preferences in relation to the giving and receiving of such aid.

MUTUAL AID CONTRACTS

In September 1998, Manningham Housing Association in a deprived neighbourhood of Bradford introduced a pilot scheme to encourage mutual aid amongst its new tenants. Applicants for social housing were requested to fill in a 'social needs audit' of their present neighbourly interactions, the tasks that they could offer their potential neighbours and the work that they would like to have undertaken for them. Having returned these audits, the Housing

Association chose 22 tenants whose offers and requests closely matched and asked them to sign a voluntary 'mutual aid' contract before handing them the tenancy. The activities involved include car maintenance and repair, computer training, babysitting and a DIY club. So far as is known, this is the first such pilot scheme in the country.

Viewed as a tool for creating jobs, this initiative offers little to policy-makers. At best, it develops self-esteem and confidence and improves the employability prospects of the residents by enabling them to develop, maintain or enhance skills that may be useful when seeking formal employment. Examined in terms of their applicability to generating alternative means of livelihood, however, these mutual aid contracts possess the potential to facilitate mutual aid in situations where it would not otherwise result. By contracting people to engage in mutual aid with others, these schemes 'kick-start' mutual aid in situations where it might not take place. These initiatives thus have a place in specific contexts. They are 'contractual' tools for facilitating mutual aid in contexts where it is perceived that such aid would be lacking. It is a *Gesellschaft*-like approach to the creation of communitarian relations. Obviously, this would not always be applicable. In this sense, they are useful in facilitating mutual aid in contexts where what we called in Chapter 6 'environmental' barriers to participation in informal work predominate. If such environmental barriers do not exist, they seem limited in their usefulness. Although they also tackle the social network barriers, it appears that their utility is more in situations where environmental barriers predominate.

This scheme could be replicated in the context of many social housing projects in both rural and urban areas. From the experience of similar initiatives such as LETS, however, issues that will require consideration are whether and how to keep accounts of this mutual aid, the nature of the quality control mechanisms to be used and ensuring that the momentum is maintained beyond the initial phase. Moreover, how to incorporate those who perceive themselves to lack human capital will need to be more fully considered. Such an initiative seems to do little to help people to tackle the human capital barriers and may even reinforce such barriers to participation by only choosing households to participate where pre-existing skills prevail or are recognised by household members to prevail. Until now, nevertheless, this project has not been evaluated. If it has proven to be successful, such an initiative could be implemented in many other localities where environmental barriers are prominent. It is a relatively simple way of developing mutual aid through a 'contractual' approach.

EMPLOYEE MUTUALS

Another idea still on the drawing board, but receiving some attention lately from the UK government, is the Employee Mutual. These are advocated as a new social institution that might act as a bridge into work for the socially excluded and provide greater security in an age of increased flexibility (Bentley and Mulgan 1996, Leadbeater and Martin 1998). These are localised bodies that the unemployed, employed and firms can voluntarily join through the payment of a weekly subscription fee. Building on the concept of LETS, members would earn points on a smartcard from their work for the Mutual that would enable them to 'buy' goods and services from it. As such, they are envisioned as 'new institutions for collective self-help' that match local demand for work with local supply. Their intention is to allow people to undertake the many one-off jobs that need doing but that they are unable to afford to do formally. Unlike present-day LETS, however, the intention is also to help employers fill vacancies and to bring together workers and businesses to meet shared needs for training.

Learning lessons from the problems of LETS in relation to social security benefit rules, the proponents of Employee Mutuals have argued from the outset for special benefit rules to be applied to members of the Mutual. These would make it easier for members to combine income from part-time or temporary work on the Mutual with benefits so as to reduce the insecurity that deters people from engaging in such organisations and making the transition from welfare to work. In return for such preferential treatment regarding the income disregard, jobless members of a Mutual would make a token contribution of 50 pence per week but would contribute at least 15 hours per week of services in kind. In return, the mutual would provide not only work but also training where necessary and childcare facilities, job searches and a job placement service, as well as job accreditation and a social life. Although still on the drawing board, the Minister for Employment in the UK has advocated the development of these new institutions. It seems likely, therefore, that pilot Employee Mutuals will be set up in the near future.

The current proposal is that central government should set up a National Council for Employee Mutuals to establish a legal and regulatory framework for the new movement. Having done this, benefit rules would need to be modified for the Employee Mutual members to allow easier transfer from welfare into work via part-time or temporary jobs and tax incentives should be given to encourage individuals and organisations to join such organisations. A series of pilot schemes would be then set up to test how different variations on the Employee Mutual model would work under different circumstances and the aim would be to create a national movement of at least 250 Employee Mutuals with half a million members by 2007. If this is to go ahead, then the suggestion here proposed is that serious consideration needs to be given at the pilot stage to evaluating two specific types of Employee Mutual: those seeking to create jobs and improve employability and those seeking to promote alternative means of livelihood.

There is little doubt so far as we are concerned that the concept of the Employee Mutual is an attempt to transfer the ideas from local currency schemes into a new vehicle that is more orientated towards job creation and improving employability than current local currency experiments. As such, for us, the Employee Mutual has been designed to shift the use of local currency towards the goal of full-employment. It attempts to create a third sector local currency initiative that is more in keeping with the goal of full-employment than the schemes already discussed above. At present, they are perhaps more envisaged in terms of being vehicles for helping employers fill vacancies and for bringing together workers and businesses to meet shared needs for training.

They can also be used, however, as a vehicle for facilitating alternative means of livelihood. By approaching the Employee Mutual whenever one has a task that needs conducting and returning the favour later by fulfilling a task for somebody else, the Employee Mutual would be akin to a LETS. In consequence, rather than view them as a new 'labour exchange' for businesses and employees to be put together, they can also just as validly be perceived as tools for bringing together individuals to engage in reciprocal exchange with each other. Given that Tessa Jowell, the UK Minister for Employment, gave a talk in mid-2001 on these new institutions, it seems likely that they may be moving off the drawing board and that pilot schemes might be set up. If this is the case, then there will be a need to consider their value not only as bridges into employment but also as facilitators of informal work.

CONCLUSIONS

Policy-making currently focuses upon the further development of more formal community-based organisations rather than micro-level informal work, the former being viewed as a more 'mature' form of community self-help (e.g. Home Office 1999). However, in chapters 5 and 6, we showed how such formal community-based organisations and groups tend to be joined predominantly by relatively affluent groups who engage in them for sociability purposes. The relatively poor, meanwhile, not only view such groups as for people other than them but also predominantly rely on micro-level informal work such as one-to-one reciprocity in order to meet their material needs. Our argument in this chapter has been that it is thus necessary to recognise how the relatively poor and marginalised currently relate to such organisations and how they presently prefer to use micro-level informal work in order to meet their needs.

The aim of this chapter, in consequence, has been to transcend the current third way approach that develops the third sector (organised community-based organisations) as a springboard into employment by outlining a new mutualism that seeks to develop the fourth sector (one-to-one reciprocity). To show how this approach can be implemented, we have here identified and documented a number of bottom-up initiatives that resonate with contemporary attitudes towards engaging in informal work. They focus upon developing one-to-one reciprocal exchange and do so using some form of tally or accounting system so that people can keep tabs on how much they are giving and receiving. The grass roots initiatives reviewed are local exchange and trading schemes (LETS), time banks, employee mutuals and mutual aid contracts.

In each case, we have displayed that if these initiatives are evaluated solely in terms of their ability to achieve full-employment then they are relatively ineffective. However, if evaluated in terms of their contribution to achieving the goal of full-engagement, their effectiveness is far greater and their role more central to its achievement. They effectively enable populations to harness their ability to help each other out and tackle many of the barriers towards participation in informal work outlined in previous chapters. Indeed, these initiatives are attractive to those pursuing full-engagement because they tackle precisely the barriers that prevent those currently least able to engage in informal work and, at the same time, provide alternative coping practices to formal employment so as to enable the capabilities of households to be enhanced.

Nevertheless, there remain many barriers that prevent the further development of these initiatives and we have reviewed how these barriers might be overcome. It needs to be restated, however, that the principal barrier to the further development of these initiatives is the hegemony of the goal of full-employment and the way in which these initiatives are consequently seen to be of only marginal relevance to poverty alleviation and building a more inclusive society. Unless this is overcome, then the importance of these complementary poverty alleviation strategies in tackling poverty will not be recognised. We thus conclude that for these initiatives to be used effectively and for the third way to become a distinct and radical departure from the past, there will need to be a shift both from an 'employment ethic' to a 'work ethic' and from a 'full-employment' to a 'full-engagement' vision.

9 The 'working citizen'

Top-down initiatives

If full-engagement is to be achieved, it is insufficient to rely on bottom-up initiatives. As the last chapter displayed, even the most prominent of these bottom-up initiatives are presently small-scale piecemeal projects that are alone incapable of harnessing informal work on a broad scale, especially for lower-income groups who do not at present widely participate in such initiatives. Just as employment creation requires top-down initiatives to be effective, so informal work needs top-down initiatives. Today, few people question the need for tax breaks, subsidies, incentives and high levels of state intervention in the market in order to generate formal jobs. Until now, however, the same has not applied to informal work. Instead, it has been viewed perhaps as some organic or natural economy. If so, this is incorrect.

Part II showed that just as formal economies require intervention, the informal sphere also needs intervention. Without such intervention, the socio-spatial disparities that exist in the ability of households to engage in informal coping practices will persist. Here, in consequence, a number of top-down initiatives are presented that could be implemented to help people to engage in informal work to help themselves and others. The first of these considered here is Citizen's Income. The second is the development of 'active community service' through either civil-ising New Deal or the creation of an Active Citizens' Credit (ACC) scheme.

CITIZEN'S INCOME

Is formal employment the most appropriate means of distributing income? Given that just 70 per cent of the UK population of working age have a job and that society is heavily polarised into multiple- and no-earner households, it appears that there is a need to reconsider whether this is the case.

One outcome of rethinking how to distribute income is the idea of a 'citizen's income'. Alternatively, this is known as a basic income, social wage, social dividend, social credit, guaranteed income, citizen's wage, citizen-ship income, existence income or universal grant. This would provide every citizen with a basic 'wage' as a social entitlement without means test or work requirement (e.g. Fitzpatrick 1999, Jordan *et al.* 2000, Van Parijis 1995, 2000a, 2000b). Eligibility is *automatic* for all citizens and *unconditional*. There would be no tests of willingness to work, or disqualification of partners doing unpaid household or caring work. It saves the enforcement and casework duties that might be required if conditions were attached and enables individuals to decide for themselves how to combine paid and unpaid activities.

With this minimum income guarantee in hand, individuals could choose to improve their well-being by engaging in employment so as to earn additional money in order to purchase goods and services, or they could instead invest their time in self-provisioning those goods and services or helping others. The aim is to give individuals and groups increased resources for taking charge of their own lives, further power over their way of life and living conditions.

The introduction of such a system of income allocation would free those presently unemployed to engage in both paid and unpaid activity to improve their well-being and as an outlet for their creative potentials and/or entrepreneurial spirit, and the 'poverty-trap' would be abolished (see Jordan *et al.* 2000, Mayo 1996, Parker and Sutherland 1998, Van Parijs 1995). Exploitative informal employment, moreover, would become a thing of the past because nobody would be obliged to engage in such work to augment income to a basic survival level (e.g. Jordan *et al.* 2000, Mayo 1996, Williams and Windebank 1998a). As Mayo (1996: 158) asserts,

> The approach of a Citizen's Income – would help to support beneficial changes in working patterns and practices by removing unemployment and poverty traps, eliminating today's black economy of undeclared earnings, raising pay levels for dirty and unsociable jobs, and encouraging work in the [unpaid] informal economy.

If implemented, it would no longer be solely the labour market that integrated people into society but also this scheme that would offer limited material security, esteem and identity.

One of its key benefits so far as proponents are concerned is that it facilitates tax-benefit integration. Tax allowances and social security benefits of all kinds could gradually be replaced by a single tax-free sum, guaranteed for every individual, irrespective of labour market or household status. This tax-benefit integration has long been a desire of left-of-centre analysts.

The origins of the idea of citizen's income have been traced to Tom Paine, Saint-Simon, Bertrand and Dora Russell, Major C.H. Douglas amongst others (Van Trier 1995). Currently, it is advocated by several eminent economists such as Atkinson (1998) and Desai (1998) as well as political philosophers such as Van Parijs (1995, 2000a, 2000b) and key social policy commentators (Jordan 1998). Among leading UK politicians, only Paddy Ashdown has given it support, and his party went into the 1992 general election with basic income in its manifesto. Recent research, however, shows that it is supported by many backbench MPs in all parties at Westminster and is on the mainstream policy agenda in Ireland (Jordan *et al*. 2000). Among the main advantages claimed for such a basic income scheme are:

- it is neutral between paid and unpaid work, giving better incentives for low-paid employment than tax credits, but allowing choice over how to combine the two;
- it treats men and women as equals, allowing them to negotiate how to share unpaid work in households (see McKay and Vanavery 2000);
- it combats exploitation, by allowing individuals to survive without relying on dangerous or demeaning work;
- it promotes economic efficiency, by ensuring that low-paid work is not given a special subsidy (as in tax credits) and hence labour power is not wastefully deployed; and
- it promotes social justice, by treating all individuals alike, and giving extra income only to those with special care needs.

Even amongst advocates of a basic income, however, it is now accepted that a fully individualised and unconditional basic income could not be introduced in one operation, if only because of the way in which it would upset the current distribution of incomes and labour supply. Instead, and particularly for the working age population, the growing consensus is that one should not proceed by cohorts or by categories, but start with a very modest (partial) basic income that would not be a full substitute for existing guaranteed minimum income provisions (Jordan *et al*. 2000).

In several nations, costing exercises have been undertaken. These have raised a series of questions. Could and should the introduction of a partial basic income be accompanied by a matching cut in both other benefits and gross wages, by reductions in other benefits only and higher taxes on wages, or by a higher uniform taxation of other benefits and labour income? Should the basic income be added to the taxable income? Should current tax exemptions on the lower range of labour income be kept as they are, reduced or abolished? Is there any short-term prospect for alternative funding (e.g. generalised social security contribution, energy tax, VAT, gross profits of enterprises)? Should the level of the partial basic income be explicitly indexed to the cost of living, to the minimum wage or to GDP per capita? Such questions are the source of a great deal of debate within the citizen's income literature and some of the issues arising will be returned to below.

On the issue of costing, nevertheless, Desai (1998) has calculated that a Citizen's Income of £50 per week (with a £12.50 supplement for those over 65) paid unconditionally to all adults over 18 would be paid for by raising the basic tax rate to 35 per cent. In his calculations, income support, jobseekers allowance, pensions and family credit are replaced by Citizen's Income. All other benefits are left unaltered. The basic tax rate is raised from 23 per cent to 35 per cent and the higher tax rate is left unaltered.

The largest gain in absolute terms, as well as in percentage terms, is for households in the lowest income decile. They gain £34.40 per week on average or 47 per cent. The next 10 per cent also gain and those in the next 70 per cent have practically no significant gain or loss. The highest 10 per cent, however, lose but the loss is £42.64 per week or around 6 per cent of their income.

Parker and Sutherland (1998) cost out a more gradualist approach of a transitional basic income of £20 per week plus a minimum wage of £4 per hour. They calculate that this would be revenue neutral if existing social security benefits were reduced by the £20, the current 20 per cent rate of income tax abolished and the standard rate raised from 23 to 26 per cent (with the latter increase partly offset by reducing Class I National Insurance contributions from 10 per cent to 9 per cent). At a higher level, however, although the threshold for higher income tax could remain virtually unchanged, a 45 per cent income tax rate would be required on taxable incomes (excluding the basic income) above £46,000.

Whatever the costs of a basic income scheme, our contention here is that such a top-down solution, although necessary, is by itself insufficient if the desire is to harness informal practices. To achieve this, such a scheme needs to be coupled with initiatives to encourage 'active citizenship'. As Gough (2000: 27) argues,

> It is not enough to pay citizens a minimum income without enabling them to participate in socially significant activities, including paid and unpaid work. Similarly, the divorce of rights from duties . . . contradicts . . . the strong link between the two . . . All persons who can, should have the right – and the duty – to contribute in some way to the common wealth.

Lipietz (1992: 99), similarly, argues that a universal basic allowance 'would be acceptable only if it meant that those who received it were prepared to show their solidarity with society, which is paying them'. The crucial issue confronting basic income, and to borrow a phrase, is that 'there are no rights without responsibilities'. Elson (1988: 29) puts this well,

> Alongside the right to a grant should be the duty, on the part of able-bodied adults, of undertaking some unpaid household work of caring and providing for those who are unable to take care of themselves. Persons already undertaking care of a young or sick or handicapped person would be exempt.

There is a groundswell of opinion that a basic income needs to be tied to some form of active citizenship. Atkinson (1998) for example argues for a 'participation income' and Lipietz (1995) for a new sector engaged in socially useful activity and comprised of 10 per cent of the labour force (the unemployment rate at the time he was writing). The idea is to tie a basic income scheme to some form of participation in society. Advocates of an unconditional basic income, however, have argued against such an approach (e.g. Gorz 1999, Jordan 1998, Jordan and Jordan 2000). For them, the result will be new forms of compulsion akin to workfare such as compulsory work in the third sector (e.g. Offe 1995, Elson 1988, Rifkin 1995).

This, however, does not necessarily follow. It is here argued that it is wholly possible to create what have variously been called, 'passports to participation', 'participation incomes' or forms of 'community service employment' (see Williams and Windebank 2001b) that provide basic incomes for active citizenship in a way that avoids compulsion.

ACTIVE COMMUNITY SERVICE

At the core of most models for security, esteem and identity is the notion of the 'working citizen'. Here, we do not challenge this model. Our only concern is to broaden out what is meant by 'work' to include informal economic activity. At present, in the social integration discourses of New Labour, the promotion of security, esteem and identity is through a model that views the 'working citizen' as somebody participating in formal employment. In this view, everything is linked to a paid job, including citizenship itself as manifested by the lack of distinction drawn between citizens' rights and workers' rights.

Here, however, and based upon a broader definition of work that encompasses informal activity, we wish to promote an alternative model for security, esteem and identity. In this view, the 'working citizen' is integrated not only through employment but also through other forms of active citizenship. To see how this new model of the 'working citizen' might operate, two policy options are here considered. The first seeks to extend the 'voluntary and community sector' of the New Deal programme so as to promote integration through active citizenship (rather than formal employment alone) and the second seeks to introduce an Active Citizens' Credits scheme that again broadens the routes to integration. Each is considered in turn.

Civil-ising New Deal

Following the lead of the US, many advanced economies are implementing workfare-type regimes as part of a general reorientation of labour market intervention towards active policies (see Lodemel and Trickey 2001, OECD 1999, 2000, Peck 2001). Workfare programmes represent a significant departure from traditional welfare systems. People are required to work in return for social assistance payments. In contrast to previous welfare and unemployment benefit programmes in which state support was passive, unconditional and entitlement-based, these new workfare regimes are conditional, work-focused and oblige participants to be active in order to receive social payments (Campbell 2000, Robinson 1998). The principal critique of these workfare regimes is that there is a compulsion element, whereby people are forced to do work that they would not otherwise wish to conduct (e.g. Peck 1996, 1999, 2001).

Our intention here is to consider a modification to New Deal that would reduce the workfare or compulsion critiques that currently blight it. It would also unleash the unemployed from the shackles that prevent them from engaging in productive and meaningful activities within their communities that they might wish to undertake but are currently prevented from carrying out. The proposal is that the 'voluntary and community' sector of the New Deal programme could be extended to allowing the unemployed to define the 'social contribution' that they wish to make. This would not only negate the workfare critiques aimed at this programme but also release the unemployed to take greater responsibility for the nature of their integration into the world of work.

Indeed, precedents already exist. In June 1998, it was announced that musicians were to be funded under the New Deal. They received a training allowance equivalent to their normal jobseekers allowance (JSA) plus a grant of £15.38 per week. The outcome was that they were paid not because they are 'available for work' but in recognition of their individual talent and contribution to society. This is a precedent to extend. It could similarly be applied to many other groups by paying them 'activity benefit' that would be higher than the 'zero-activity benefit' level.

To achieve this, individuals could be empowered to stake a claim under the 'voluntary and community sector' of New Deal concerning their contribution to society. The precise scope of work that might be acceptable and the problems involved in deciding on the breadth of such activity are dealt with below but would certainly include caring activities and organising community groups. Hence, somebody who was principal carer of a young pre-school child or an elderly dependent person would have this essential work recognised under the 'community and voluntary sector' of New Deal and they would be paid an activity benefit for doing this work. Similarly, those organising and running community groups such as LETS and credit unions would again receive an activity benefit paid at a higher rate than the zero-activity benefit level in recognition of their contribution to community revitalisation.

More widely, if such an approach provided some leeway on what was considered acceptable activity, then this could both encourage and facilitate 'social entrepreneurship', especially in deprived neighbourhoods (Leadbeater 1999). After all, it is the individuals living in these neighbourhoods who are best positioned to understand the needs and creative desires of these communities and how these are not being met at present. Such an initiative could mobilise this 'local knowledge' by allowing individuals to test out their ideas on 'gaps in the market' or 'unmet needs'. By adopting a more flexible approach towards defining acceptable activity for which an 'activity benefit' would be paid, therefore, such a policy would facilitate:

- the development of new, locally rooted micro-businesses serving local and regional markets;
- the formalisation of informal economic activities;
- the building of community capacity through the mobilisation of local communities and their neighbourhood resources; and
- the encouragement of self-help through the development of local solutions to locally identified problems.

An example of a similar initiative is to be found in Australia. Here, there has been some recognition that various activities are socially legitimate for those who are claiming out-of-work benefits. It has been understood, for example, that care work should be accredited. Both lone parents, and one parent in a couple, can claim a 'parenting allowance'. This is income-tested on one's own and one's partner's income, and is payable to parents with children aged up to 16 years old (Hirsch 1999).

Although this policy proposal would start to conquer the workfare critiques associated with active labour market policies such as the New Deals, a key problem that this proposal fails to modify is the meaning of a 'working citizen' amongst those who are not eligible for New Deal. The likely outcome is that it would introduce a 'dual society'. Those eligible to choose the contribution that they wish to make to their communities are only those who are unemployed. The likelihood, therefore, is that the new forms of work undertaken would be seen as a second-rate and second-class 'economy' for those excluded from the formal labour market. Below, therefore, we present a more comprehensive top-down strategy for facilitating a full-engagement society that is more inclusive in terms of the groups who could be mobilised to renegotiate their contribution to society.

Active Citizens' Credits (ACC)

The conventional contract between the state and out-of-work households offers income in exchange for a duty to search for employment if one is able. Only those considered inactive such as because they are sick and disabled, rather than those engaged in alternative activity, are exempted from this obligation. The biggest exception to this rule in the UK (but no longer in the US) has been lone parents, who have effectively been allowed to opt for parenting as an alternative to employment.

Here, however, a scheme is proposed that could not only recognise the contributions of work beyond employment but also reward those individuals who engage in such endeavour. This scheme is based on the notion of accrediting active citizenship. Drawing upon the ideas for Citizens' Service (Briscoe 1995, Hirsch 1999, McCormick 1994) and a participation income (Atkinson

1998), we here advocate the development of 'Active Citizens' Credits' (ACC). The intention of this ACC scheme is to provide a means of recording, storing and rewarding participation in caring and other work conducted for the good of their community. Under this non-compulsory scheme, individuals would engage in a self-designed portfolio of work of their choosing for which they would be reimbursed.

It would be non-compulsory in the sense that individuals could freely choose whether or not to participate. Participation would not be obligatory. It would also be grounded in the idea that the individuals engaging in such work would decide for themselves the portfolio of work that they wished to carry out. Put another way, it would not be based on the notion that individuals should be delegated tasks by institutions on the basis that these institutions believe that those tasks need to be undertaken for the good of the community. Instead, individuals would be active participants in deciding what they wished to do, although facilitators would be employed in order to help them arrive at their decision, if individuals wished to draw upon this resource.

The goals behind such a proposal are:

- to recompense and value work that currently goes unrecognised and unvalued;
- to encourage active citizenship without recourse to compulsion;
- to harness informal work;
- to create a 'full-engagement' society by enabling people who wish to make a particular 'social contribution' to do so;
- to incorporate the multi-dimensionality of social inclusion and exclusion into policy-making; and
- to tackle social exclusion and promote social cohesion through means other than merely insertion into employment.

The idea that such a scheme should be developed to encourage individuals to engage in freely chosen work to benefit their community is perhaps uncontroversial.

The major issue that needs to be dealt with, however, is how to reward people for their active citizenship. Our proposal is that this can be achieved by extending the tax credit approach that is emerging in many advanced economies (e.g. Liebman 1998, Meadows 1997, Millar and Hole 1998) and was introduced in the UK in October 1999 (Bennett and Hirsch 2001, HM Treasury 1998).

In the UK, tax credits have been at the heart of welfare state reform. A clear signal that this was to be the case followed Frank Field's resignation in July 1998 as special Minister for Welfare Reform, when this function moved into HM Treasury and a key architect of the Working Families Tax Credit (WFTC) was given responsibility for Welfare State Reform. Under WFTC, and continuing the Family Credit system, working families with children were prioritised in that a parent working over 16 hours per week was effectively guaranteed a minimum income (HM Treasury 1998). Therefore, families without children, single people, part-timers working less than 16 hours per week and the unemployed were excluded from WFTC, despite such groups facing the same 'poverty' and 'unemployment' traps as working families with children.

This was to change in the April 2002 Budget statement of the UK Chancellor of the Exchequer when the tax credit system was extended to incorporate single people, couples aged over 25 years old without children, lone parents and pensioners. The intention behind this rolling out of the WFTC to a 'Working' tax credit, or what we prefer to call an 'Employment' tax credit, is to move towards a fully integrated tax-benefit system using a tax credit approach. How, therefore, must the tax credit approach be extended in the future in order to create this integrated tax/benefit system?

Presently, the unemployed and part-time employed people are not included. If these groups were paid a guaranteed minimum income, then they would be paid this for engaging in employment for fewer hours or not at all. Although exponents of an unconditional citizen's income might support this (e.g. Jordan 1998, Jordan and Jordan 2000), we here argue that such tax credits should not be paid for doing nothing or doing less. Those able to do so should be required to make a 'full' contribution to society to warrant their tax credit/guaranteed minimum income. How could this be achieved?

Our proposal is to give everybody who wishes to receive the guaranteed minimum income the opportunity to do so. To achieve this requires a further extension of tax credit in the manner outlined in Table 9.1. Following the 2002 Budget Statement in the UK, this is now well under way compared with our earlier discussions of this issue. As we previously outlined (see Williams and Windebank 1999b, 2001b), our advocacy of an extension of the WFTC to an Employment tax credit or what might be called Labour Market Participation tax credit has now been introduced under the banner of a 'Working Tax Credit'.

In addition, our earlier proposal that those reaching retirement age should receive a Pensioner tax credit has also now been introduced from April 2002. The result is that several additional tax credits now remain that need to be introduced in order to create a fully comprehensive tax credit system.

First, and as outlined in Table 9.1, there is the Disability and/or Sickness tax credit for those absent from work on the grounds of sickness, injury or disability, which might be implemented in a very similar manner to the current Disability Living Allowance. Again, this is an initiative that has received some attention in recent years but has so far made little progress not least due to opposition from disability pressure groups. Second, and to incorporate those engaged in post-compulsory training and education in order to feed the 'knowledge economy', a Training tax credit would be introduced for those engaged in full-time education or training.

Table 9.1 A comprehensive tax credit framework for the UK

Tax credits	Qualifying condition
Employment tax credit	Working as a full-time employee or self-employed
Disability and sickness tax credit	Absence from work on grounds of sickness, injury or disability
Pensioner tax credit	Reaching pension age
Training tax credit	Engaging in approved forms of full-time education or training
Active Citizens Tax Credits, comprising:	Participating in caring activities and other work for the good of the community
Parents tax credit	Principal full-time carer of a pre-school child
Carers tax credit	Principal full-time carer of a dependent adult
Community worker tax credit	Engaging in full-time work for the good of the community

Even if both tax credits were introduced, however, some would still not meet any of the qualifying conditions for the Employment, Disability and sickness, Pensioner or Training tax credits (i.e. full-time employment, absence from work on grounds of sickness, injury or disability, pension age, in full-time education). This is because they are unemployed and cannot find a job, or because they are engaged in employment or training only on a part-time basis. Our proposal, therefore, is that a final tax credit could be introduced that would provide these people with the opportunity to receive a guaranteed minimum income in return for engagement in active citizenship.

This final form of tax credit we have called Active Citizens' Credits. No compulsion is attached to participation in this scheme. Indeed, to overcome the compulsion criticism, which has so heavily been targeted at the New Deal, the difference between this proposal and the New Deal is that opportunities to claim Active Citizens' Credits could be created from the bottom-up in three ways:

- by employing people on initiatives set up by central or local government and third sector agencies that have identified particular realms in which needs are not being met;
- by individuals who in partnership with ACC facilitators (e.g. community development officers, NGOs, voluntary sector bodies and agencies, TECs, FE colleges) are helped to design their own ACC contract; and
- by individuals who autonomously create their own ACC portfolio for validation and scrutiny.

That is, and unlike at present, individuals would be given the option of designing their own ACC portfolio of work rather than having to rely on the state, voluntary organisations and/or market to find them work. It would also provide individuals with a sense of ownership of their work due to its nature being autonomously chosen by the person. Consequently, individuals would be able to stake a claim for what constitutes their contribution to economic and social life and/or to create a portfolio of activity to make such a contribution.

Hence, being unemployed but available for work would not be a qualifying condition for tax credits, since such people could be making some social contribution. Similarly, those employed for less than 16 hours would not automatically receive a tax credit. They could only do so by 'topping-up' their part-time employment via either education or participation in this final category in order to qualify for a guaranteed minimum income. Of course, the unemployed who did not wish to do so would not receive this 'activity benefit' and would instead receive a 'zero-activity benefit' that may or may not be at a lower level than an Active Citizens tax credit. The advantage of this approach is that it enables a shift from the present system where 'for long-term unemployed and disabled people, there is a positive discouragement to engage in socially useful activities, for fear of losing benefit' (Hirsch 1999: 8) to a situation where such activity would be encouraged.

What type of work would be undertaken under ACC? The scope of activities included under ACC is here left open for discussion and is very much a matter for political debate. At a minimum, however, and as Table 9.1 displays, we argue that it should include not only caring activity for young, elderly or disabled dependants but also any service activity that is undertaken by individuals for the benefit of their communities, such as organising community groups. Indeed, this is the reason why Active Citizens' Credits have been subdivided into three further types of tax credit: parents tax credits, carers tax credits and community worker tax credits (see Table 9.1). Here, therefore, we outline the nature of each of these tax credits in turn.

Parents tax credit

As a society, pensioners receive a guaranteed minimum income because they have worked, contributed to society and then are out of the labour market. Yet there is no equivalent recognition for women, or more unusually men, when out of the labour market because they are caring for young children. The assumption that married women are cohabiting and can afford to stay at home because of a 'male breadwinner' is no longer valid.

In 2000, Harriet Harman, chair of the UK Childcare Commission, set up to advise the government on how it can best supply childcare in the future, called for mothers to be paid to stay at home to look after young children until they are two and a half years old. She recommended a payment of £150 per week or £7,800 per year. If implemented, this would overcome many of the anomalies that have arisen with the introduction of the WFTC, particularly the childcare tax credit. For example, a parent can currently claim childcare tax credit for a registered childminder. However, such credit cannot be claimed if they provide the care informally, despite the fact that such kin-based care is perceived to be usually of a much higher quality than when it is collectivised in crèches and nurseries (e.g. Mauthner *et al.* 2001, Windebank 1996, 1999). By redefining mothers, and sometimes fathers, as engaged in parental care for which they would be given a parents tax credit, the present anomaly would be overcome and at the same time, their activity would be recognised as work. Unlike at present, a parent would receive credit for their activity rather than benefit for their inactive status. The outcome would be to incorporate the 'care ethic' into the 'work ethic', something scholars such as Lister (1997) have propounded.

For this to be achieved, however, recognition is first required that work is more than employment and that the 'working citizen' means more than being a formal employee. Presently, and based on the view that insertion into formal employment is the principal means of social inclusion, the contorted reality prevails whereby one can be paid to look after others' children but not one's own (Hills 1998). A tax credit for mothers, and for fathers where they have primary parental responsibility for childcare, would resolve this anomaly.

If implemented, there is little doubt that this would be popular. The childcare of choice for most mothers is themselves or their partners (e.g. Mauthner *et al.* 2001, Windebank 1999). Old-style social democrats in the UK, however, have tended to view 'formal' childcare, including childminders, playgroups and nurseries, as preferable, while the New Right has cornered the option of parental childcare. However, there is no reason why this ideological division should remain in third way thought. Indeed, the Swedish Left has been for a long time supportive of home childcare. A parental tax credit, in consequence, would resolve current anomalies and in a manner that the population at large would find desirable.

It might also resolve the inequalities that will arise with the recent introduction of unpaid 'parental leave' in the UK. Although unpaid parental leave of 13 weeks is now available to parents (18 weeks for parents with registered disabled children), there is recognition that take-up will be limited to the more affluent who can afford such an option. According to the Low Pay Unit, however, a new parental leave tax credit for low- and middle-income families would add as little as £26 million to the £5 billion annual bill for WFTC (see Ward 1999). Seen in this light, the introduction of parental tax credit is a means of resolving not only current anomalies arising with the childcare tax credit but also the problems with the provision of unpaid parental leave.

Before leaving this parental tax credit, it needs to be recognised that paying parents for their childcare function is not new. This proposal simply extends principles already enshrined in legislation. National Insurance policy, for example, already recognises the valuable contribution of parents in rearing children on a full-time basis in the sense that it pays National Insurance contributions for the period spent out of the labour force engaged in such work. The aim here is simply to extend this principle by further formalising the recognition and value attached to such work so as to overcome the present anomalies in the tax credit system.

Carers tax credit

Caring, however, is not limited to parents who look after children. A large number of people also care for other dependants such as the elderly, disabled, sick or injured as well as grandparents looking after children. In the UK, some 11 per cent of the population are carers, only about a quarter of whom receive Carers Allowance (Desai 1998). At the EU level, meanwhile, some 6 per cent of EU citizens are engaged in unpaid informal care of sick or disabled adults and older people (in the same household or outside). On the whole, and reflecting the gender division of caring, these are usually women. Women are twice as likely as men to be caring for sick or disabled adults or older people on a daily basis (8 per cent compared with 4 per cent). The average number of hours spent on this care is 21 hours per week (European Commission 2000a). Moreover, just over one in five carers say that caring responsibilities prevent them from undertaking the amount of paid work that they would otherwise do and they are overwhelmingly women in the twenties and thirties (European Commission 2000a).

The introduction of a carers tax credit would resolve many of the problems confronting this segment of the population. It still remains the case that primarily seeking 'inclusion through employment' results under New Deal in many who are principal carers being 'helped' to return to employment rather than their valuable caring work being valued. By recognising these people as active citizens and paying them for their caring work, many of the problems that they confront when they are treated as inactive and encouraged to enter formal employment would be resolved (Hall 1999).

Community worker tax credit

In recent years, there have been many calls in government documents for greater 'civic engagement', 'self-help' and 'community involvement' (e.g. Home Office 1999, Social Exclusion Unit 1998, 2000). Presently, however, the government has no way of rewarding those who engage in such activity. Indeed, it is often the case that the work of those who make a valuable contribution to their community is not recognised. Under New Deal, for example, somebody who may have set up and is running a community-based group such as a LETS, but is officially viewed as unemployed, does not have the valuable contribution that they are making to their community recognised. Instead, they are encouraged to give up this work and to insert themselves into formal employment under New Deal. However, inserting people into formal employment as these individuals might testify is not the sole means of pursuing social integration and inclusion.

By introducing a community worker tax credit, the contributions of those who are economically active and making a significant contribution to their community (e.g. unemployed individuals who set up and run community-based initiatives such as LETS and credit unions) would be recognised. Presently, their work is unrecognised and, indeed, is sometimes actively discouraged by a work culture based on the notion that everybody, if not engaged in, should be seeking, conventional employment. However, such activity is arguably at least as important as tasks undertaken through conventional employment, even if they are presently not recognised or recompensed on an equal basis with employment. The result of a Community Worker tax credit would be that many people who currently find themselves pressurised to give up such meaningful productive activity and seek employment would be released to devote themselves to such work.

In so doing, the current proposals to add a third prong to the welfare equation in the form of the third sector would be given a significant boost. One might envisage those engaged in activities such as organising pre-school playgroups, community-based rural transport initiatives, LETS and credit unions being paid under this tax credit.

Towards the 'working citizen': implementing Active Citizens' Credits

Similar to the basic income scheme (Atkinson 1998, Jordan 1998), it is likely that paying everybody a full tax credit at rates similar to WFTC whatever their social contribution is unfeasible on the grounds of cost. Three alternatives thus exist. First, one could pay ACCs according to total household income, which is indeed what has started to occur from April 2002 onwards in the UK with the child tax credit paid direct to the principal carer, breaking with a long tradition of individual taxation. This is fairer in terms of socio-economic justice and by paying it to the principal carer, the hope is that this will avoid a transfer from the purse to the wallet. However, although it is possible to avoid the latter transfer with child tax credit, it is not possible with many of the other tax credits.

A second option, in consequence, is to pay different levels of tax credit according to the qualifying condition met by the individual. Thus, pensioners, those in full-time education and the various types of ACC participant (e.g. parents, carers and community workers) might receive less than the full credit available to formal employees. This activity-related tax credit system, however, would need to be at a level to exempt such groups from the need to claim additional means-tested benefits if it is to represent a fully integrated tax-benefit regime. Indeed, this has started to occur in that working families, lone parents, couples without children and pensioners all receive different levels of tax credit. A principal problem, nevertheless, is that if applied to Employment tax credits compared with ACCs it would perpetuate the valuing of employment over other forms of work. Nonetheless, at least it would commence the process of more fully recognising and valuing the work that is currently unvalued in society.

A third and final option is to limit ACCs in the first instance to particular localities and/or social groups. For example, one might envisage ACCs being made available at the outset only to the long-term unemployed, the early retired or deprived neighbourhoods. If this were to be implemented, however, there is the inherent danger that it might be classified as signifying the emergence of a second-class form of employment and the creation of a dual society. Nevertheless, the introduction of ACCs in various stages incorporating different population groups at each juncture, just as with Employment tax credits, would enable those least likely to be in the labour market to be targeted in the first instance. After all, the current reforms of the tax-benefit system centred on insertion into employment are aimed at improving incentives for low earners to become 'citizen workers'. For many citizens, however, especially those living in deprived communities with structural economic problems, lack of labour demand prevents these reforms from having much practical effect. The expansion of what constitutes work in the way described above, however, would open up many more opportunities for these populations than are currently available.

A further issue, whichever option is adopted, concerns whether the level at which all of these tax credits are paid (not just ACCs) should be spatially variable. In Chapter 4, we revealed significant spatial variations in coping capabilities and these appear to be the result of flat-rate payments for social security, pensions and so forth that do not recognise the spatial variations in the cost of living. Any new form of social payment, therefore, will need to consider whether it is feasible to respond to these disparities by seeking to *regionally differentiate* the amount of money paid.

To see why it is important to consider regionally differentiating tax-credit payments (whether the qualifying condition is employment, reaching pension age, training or active citizenship), Table 9.2 reveals how expenditure on WFTC is currently unevenly distributed between regions. Take, for instance, London. According to the 1998 Index of Local Deprivation (that combines a selection of indicators covering crime, housing, unemployment and health to capture the extent and location of multiple deprivation), 13 of the 20 most deprived areas are in London and almost half (48.5 per cent) of London boroughs are in the worst 50 local authority districts (Cabinet Office 1999). Yet despite the concentration in London of wards suffering from multiple deprivation, just 1.8 per cent of its population receive WFTC compared with 2.6 per cent nationally. One principal reason for this is that relatively high wage rates in London exempt a large proportion of its population from qualifying for WFTC, despite the much higher cost of living in this city (Reward Group 1999). Such a situation clearly reveals the need for a regional differentiation of WFTC to be considered.

This need to at least consider a regional differentiation in the level of tax credit payments is further revealed when the national minimum wage

Table 9.2 Regional distribution of recipients and expenditure on Working Families Tax Credit (WFTC)

Government office region	WFTC recipients		Expenditure (£m)
	000s	*Percentage of regional population*	
North East	85	3.3	275
North West & Merseyside	200	2.9	625
Yorkshire & Humber	130	2.6	425
East Midlands	105	2.5	350
West Midlands	125	2.3	525
Eastern	115	2.2	400
London	125	1.8	475
South East	110	1.4	500
South West	115	2.4	350
Total	1,110	2.6	3,925

Source: Cabinet Office (1999: 75–6).

(NMW) is considered. Table 9.3 reveals that London, the South East and the Eastern regions have a lower proportion of employees eligible for the NMW than other regions. Despite 3.1 per cent of all employees nationally being eligible, eligibility levels are just 0.9 per cent of employees in London, 2.0 per cent in the Eastern region and 2.5 per cent in the South East. This is again because the NMW takes no account of the regional variations in the cost of living. As one of the few analyses of the regional impacts of these national-level work and welfare policies reveals, if the regional cost-of-living indices produced by the Reward Group are used to deflate the adult minimum

Table 9.3 Regional distribution of recipients of the National Minimum Wage (NMW)

Region	Employees (000s)	NMW eligibility as percentage of all employees
North East	40	4.5
North West/Merseyside	120	4.7
Yorkshire & Humber	70	3.6
East Midlands	60	3.7
West Midlands	70	3.3
Eastern	40	2.0
London	30	0.9
South East	80	2.5
South West	90	4.9
Wales	50	5.2
Scotland	40	2.0
Northern Ireland	20	3.3
United Kingdom	710	3.1

Source: Cabinet Office (1999: 79).

wage (£3.60), in real terms the national minimum wage would vary from £3.18 in Northern Ireland to £4.09 in Greater London (Sunley and Martin 2001). Put another way, if the national minimum wage is £3.60 in Northern Ireland, it would have to be £4.43 in Greater London to have the same value in real terms. If such differentials were introduced, then many more in these southern regions, for example, who currently find themselves on *real* wages that are as low, if not lower, than those living elsewhere, would become eligible to receive the NMW.

It appears therefore that any introduction of a tax credit approach will require consideration to be given to regionally differentiating payments in order to take account of the regional variations in the cost of living.

Another issue concerns whether there should be a *time limit* attached to ACC. This would have to be differentiated according to the work involved. For childcarers, for example, it might be limited to the first five years of each child's life (i.e. until all children are of compulsory school age), whilst for eldercare and for community workers, there might be no time limit. This, however, depends upon whether one is pursuing a goal of full-employment or full-engagement. For adherents to the goal of full-employment, who view such work as a springboard into employment, time limits would doubtless be introduced into this sector in a bid to make the scheme self-financing. However, for those who recognise that such work can never be self-financing but nevertheless provides a complementary means of meeting essential needs, no time limits would be envisaged. The debate on time limits, therefore, revolves around whether such work should be seen primarily as a spring-board into formal employment or as a complementary means of livelihood (see Chanan 1999).

Problems and prospects for an ACC scheme

A potential problem with this ACC scheme might be that the policing of such a strategy would be overburdensome and impinge on private life (e.g. Jordan and Jordan 2000, Jordan *et al*. 2000). However, there is currently a major burden of administering a system to ensure that people do not work whilst claiming benefit and this has significant impacts on the willingness of the unemployed to help anybody else in case it is misconstrued, as revealed in chapters 5 and 6 (see also Williams 2001b). Our proposal is that all of this energy currently put into stopping people from working should be harnessed to ensure that people are working.

It could be also argued that ACC might compete with conventional employment and take away jobs from the formal labour market. This, however, is not the intention of the ACC scheme. By focusing upon parent-ing work, caring and community work, there is of course the issue that this work could be provided by the formal labour market. However, so far as parenting is concerned, it is responding to the desire of most parents to do this work for themselves. In the realm of community work, meanwhile, there are currently numerous needs that are unmet or which are being met on an unpaid one-to-one basis and it is these that should be targeted by ACC. It is intervening in the spheres where the market shows no signs of imprinting itself.

On another level, this ACC scheme might be seen by many of a radical European social democrat or communitarian ecocentrist persuasion as not going far enough. For these analysts, the objective is to facilitate a multi-activity society, which is usually interpreted to mean individuals engaging in multi-activities. This proposal only rewards citizens for engaging in a single form of work so does not encourage multi-activity amongst individuals. However, in the sense that it recognises and values work beyond employment, it will doubtless be seen as a first step towards a multi-activity society. One interpretation of this ACC scheme, therefore, is that it represents a tentative preliminary step towards the establishment of a multi-activity society in which individuals engage in multiple activities for which they are equally rewarded.

A further problem is that it only recognises work beyond employment in conventional money terms. It does not challenge the hegemony of the idea that 'money is a measure of all things'. At present, it is difficult to envisage how this might be achieved. Perhaps, and as studies of alternative currencies have revealed, the issue is not so much using money to recognise work beyond employment, but the nature of the money that is used. Conventional currencies, where scarcity is inherent, produce different social relations of exchange and measures of value from community currencies where scarcity is not a feature. For some, in consequence, reimbursement for work beyond employment might be in the form of a wider range of 'monetary rewards' than is discussed above. As Beck (2000: 130) asserts when proposing the closely related idea of developing 'civil labour':

> Civil labour is not paid, but is rewarded both materially and non-materially through civic money, qualifications, pension entitlement and 'favour credits' (for example, entitling someone engaged in civil labour to send their child to a crèche free of charge).

Rather than using just conventional money to reward work beyond employment, one could thus also envisage using time currencies, LETS, pension entitlements and qualifications. To take just one example, in the UK, the government might reward such civic activity by paying into the stakeholder pension the maximum amount currently allowed of £3,600 per annum.

Whatever approach is adopted, however, the key issue is that if a full-engagement society is to be engendered, then there will need to be a means through which active citizenship can be recorded, stored and rewarded. As shown, most people are unwilling to engage in reciprocal exchanges beyond kin unless there is some form of tally system through which their contributions can be recorded, stored and rewarded. This proposal attempts to work with the grain of current desires to provide an integrated tax-benefit system through a comprehensive tax credit system. By rewarding active citizens in the form of tax credits, a comprehensive and integrated tax-benefit system could be introduced that also recognised and valued work beyond employment. The result would be the creation of a society founded upon the principle of full-engagement without a radical policy overhaul.

CONCLUSIONS

In sum, this chapter has set out some top-down policies to harness work beyond employment. Rather than advocate an unconditional basic income, we have here explored various approaches that could be used to implement a citizen's income conditional upon people being 'worker citizens'. First, we have proposed the extension of the 'voluntary and community sector' of the New Deal programme by giving the unemployed greater freedom to define the social contribution that they wish to make. The problem, however, is that this limits engagement in work beyond employment to those who qualify for the New Deal and has the inherent danger of creating a dual society. Second, therefore, and in order to open up such opportunities to a wider range of people, we have proposed a more comprehensive option. This involves the development of an Active Citizens' Credit (ACC) scheme that would seek to provide a means by which citizens can engage in various forms of work beyond employment of their choosing for which they would be reimbursed. Three types of ACC scheme have here been proposed: parents tax credit, carers tax credit and community worker tax credit. If implemented, these tax credits would harness work beyond employment, facilitate the emergence of a full-engagement society and provide the basis for a fully comprehensive and integrated tax-benefit system founded upon the 'citizen worker' model.

Given that the needs of the poor and unemployed are currently being met so inadequately by the conventional job creation model alone, it is obvious that new models of social integration are required. It is to be hoped that this tentative proposal will help engender more debate than has so far been the case on how this might be achieved. What is certain is that the approach of inclusion through formal employment alone is no longer feasible or desirable. It is time that new proposals were more fully considered. This ACC scheme is but one possibility that needs to be considered.

Conclusions

This book has shown that discourses on poverty and social exclusion beyond the 'Old Left' (old-style social democracy or the 'first way') and the 'New Right' (neo-liberalism or the 'second way') are not confined to New Labour's approach of seeking social integration through paid employment. Here, we have articulated another discourse that is starting to emerge. Rejecting the infatuation of New Labour with inserting people into employment (e.g. Beck 2000, Giddens 1998, 2000, Jordan 1998, Levitas 1998, Lister 1998, 2000), this alternative third way argues that poverty can only be alleviated and a more inclusive society developed by giving greater value and recognition to work beyond employment. Our aim here has been to provide a clear elucidation of this new discourse and to explore its implications for tackling poverty and social exclusion.

To articulate and develop this alternative third way, Part I of this book articulated the rationales for recognising and valuing work beyond employment, Part II developed a means of examining poverty that resonates with this new stance and Part III investigated the policy implications of adopting this emerging approach. In this concluding chapter, we pull together the findings. In so doing, our aim is to start to chart a new third way approach towards poverty beyond not only old-style social democracy and neo-liberal thought but also New Labour's employment-centred approach.

RATIONALES FOR A THIRD WAY THAT RECOGNISES AND VALUES WORK BEYOND EMPLOYMENT

In recent years, New Labour has taken some interest in harnessing work beyond employment. Until now, however, this interest has been limited to the realm of welfare provision and more organised voluntary groups. The third sector has been seen as a third prong in the welfare equation that can complement the public and private spheres in meeting welfare needs. Little attention, however, has been paid to harnessing the full range of work beyond employment as an 'economic' policy tool. The steadfast view of New Labour is that informalisation is something to which they are opposed and that insertion into formal employment remains the best route out of poverty. In Part I of this book, however, this standpoint has been challenged.

To do this, Chapter 1 outlined the 'employment problem' confronting advanced economies and evaluated critically the use of employment as the principal route out of poverty. First, this highlighted the significant gap between current employment participation rates and a full-employment scenario, despite all of the efforts in advanced economies to achieve this goal. Even in the UK, which is one of the advanced economies with the highest employment participation rates, some 30 per cent of the working-age population remain outside of the formal labour market. These people, moreover, are not the wives of employed spouses. Instead, 15.4 per cent of the UK population are clustered together in jobless households and these households comprise 91 per cent of all households living below the poverty line. This large segment of the population excluded from the formal labour market and living in poverty, therefore, can no longer be ignored. When it is recognised that employment participation rates are making no significant progress in getting anywhere close to a state of full-employment, the issue arises of whether such a heavy emphasis should be put on job creation alone to solve the problem of poverty.

This is further reinforced by the recognition that 'full-employment' has now been redefined. At the Lisbon Summit of the European Council, it was decided that full-employment means achieving the same employment participation rate as our competitor trading blocs, namely North America. The intention, in consequence, is to achieve an employment participation rate in the EU of 70 per cent by 2010. What, therefore, is to be done with the 30 per cent of the working-age population who it is now accepted will not become integrated into the formal labour market? For us, this is a key question that requires an answer and is a principal reason for rethinking the goal of 'full-employment' and the associated idea that employment is the best route out of poverty.

A second reason for rethinking both the goal of full-employment and the idea that insertion into employment is the main route out of poverty relates to the desirability of this approach. The emergence of the working poor and the increasing rejection of an 'employment-centred' lifestyle add further weight to the need to rethink the trajectory of society. Evidence has been here presented that displays not only the emergence of the working poor but also how it is now the income from employment rather than employment itself that provides people with the motivation to enter the formal labour market. Consequently, if full-employment appears extremely doubtful and there is a decentring of the importance of employment in many people's lives, the question becomes one of whether employment creation alone should be used as the principal route out of poverty. In Chapter 1, we raised the idea that additional routes could be constructed based on recognising and valuing work beyond employment.

The need to build such supplementary bridges out of poverty was shown to be all the more important in Chapter 2 where it was revealed that work beyond employment is not some residual and diminishing sphere but, rather, is extensive and growing. The dominant narrative of economic development is of the natural and inevitable formalisation of societies as they become more 'advanced'. However, this chapter revealed that employment is by no means the dominant or even a growing means of performing work, at least measured in terms of the time spent on economic activity. The result is that questions have been raised over whether job creation alone should continue to be seen as the royal road to poverty alleviation. Over half of all working time is spent conducting non-market work and time spent in the informal sphere is growing relative to time spent on market-orientated activity. The question that thus arises is whether greater emphasis needs to be given to harnessing the other half of the economy by enabling people to engage in informal modes of production. This is made all the more pertinent when the evidence on participation in informal work is reviewed from around the advanced economies. Those excluded from formal employment are also less able to engage in such activity than their employed counterparts, meaning that informal work consolidates, rather than reduces, the socio-spatial disparities produced by the market.

In Chapter 3, therefore, the various approaches that can be adopted towards the informal sphere and the implications of pursuing each of them have been reviewed. These options are:

- to enable deprived populations to rely less on informal economic activities by giving them access to formal-sector provision through either employment creation and/or higher benefits;
- to allow the situation to continue as it is by adopting a laissez-faire approach towards informal work; or
- to swim with the tide of these structural changes and harness such work.

The first option of more fully formalising work and welfare, we have shown, is both impractical and undesirable. It is impractical because informal modes of production are deeply embedded in everyday life and the evidence points towards an informalisation rather than formalisation of work and welfare. It is undesirable because this mode of production is often people's preferred means of conducting many activities and a key ingredient of the social cement that binds communities together. A laissez-faire policy, meanwhile, has been shown to result in numerous negative consequences. It has been shown that lower-income populations engage in less informal work than the employed, and this laissez-faire approach has been revealed merely to leave those excluded from formal jobs also bereft of the ability to help themselves through the informal sphere. Here, therefore, it has been suggested that the only viable policy option is to swim with the tide of these structural shifts and to harness such work.

To discern the contrasting discourses on harnessing work beyond employment, two questions have been posed. What type of work beyond employment are we seeking to cultivate? And for what purpose are we seeking to harness such work? This has enabled four contrasting discourses on harnessing informal work to be identified. First, there are some, harking back to old-style social democracy, whose approach towards poverty alleviation is grounded in seeking a return towards full-employment. For them, the value of informal work is solely as a springboard into employment and organised third sector associations are focused upon to deliver this goal. The role of informal work is thus relevant solely in terms of its function of providing alternative routes into employment. Second and third, there are those who view informal work as complementary to employment creation in creating a mixed economy of welfare and/or work, and fourth and finally, there are those who view informal work as an alternative to employment. Each of these approaches has been outlined in turn.

This has revealed that New Labour's approach accepts the positive role such activity could play in providing a third prong in the welfare equation to complement the public and private spheres, but in the economic sphere, it seeks only to harness the informal sphere as a springboard into employment. Adopting the arguments of the radical European social democrats and communitarian ecocentrists, however, we have argued that there is a need to transcend this false dichotomy between the economic and social, and to consider the role of harnessing informal economic activity in the 'economic' realm. To explain why this is the case Part II has investigated poverty in contemporary England.

EXAMINING POVERTY IN CONTEMPORARY ENGLAND

Until now, little information has been available on either the contemporary extent and nature of household coping capabilities and practices in England or the barriers to participation that are confronted by those seeking to enhance their capabilities through the informal sphere. In Part II, therefore, we reported the results of a 'baseline' study that investigates these capabilities and practices in lower- and higher-income English urban and rural areas.

To do this, Chapter 4 began by examining poverty from the perspective of the 'outputs' of households rather than the 'inputs' into them. Drawing upon Sen's (1998) work on capabilities, this investigated household coping capabilities so as to provide an approach towards understanding poverty that allows the contributions of informal modes of production to be more fully integrated into accounts of poverty and social exclusion. Chapter 5 then explored the coping practices adopted by households. This revealed the socio-spatial variations in the extent to which the informal sphere is currently employed. Chapter 6 thus turned its attention towards identifying the barriers to participation in informal economic activity. It also identified the complex and dynamic interrelations between greater insertion into one sphere and its implications for participation in other spheres of activity. This high-lighted how insertion into employment, especially low-paid employment, results in a reduced capability to participate in the informal sphere and its consequences for overall household coping capabilities. Similarly, it displayed how participation in informal work is largely dependent upon having sufficient income. How this conundrum could be tackled was the subject of Part III.

TACKLING POVERTY: POLICY INITIATIVES FOR A THIRD WAY THAT RECOGNISES AND VALUES WORK BEYOND EMPLOYMENT

Having shown that informal work is widely used to perform necessary work and that most households wish to engage in more than they presently perform, but that significant barriers prevent further participation in such activity, Part III of this book has proposed a range of initiatives to cultivate participation. Before outlining these strategies, nevertheless, a clear identi-fication of the goal of such policies has been outlined.

Conventionally, the intention has been to alleviate poverty by inserting the unemployed into the formal labour market with the aim of achiev-ing full-employment. In Chapter 7, however, it has been argued that to put all of one's eggs into this basket is unrealistic. Given the importance of informal modes of production in the coping strategies of households and the apparent macroeconomic shift towards such work, it appears more feasible to pursue a policy of 'full-engagement' based on a wider conceptualisation of work.

By 'full-engagement' is meant the provision of sufficient work (both formal and informal) and income so as to give citizens the ability to satisfy both their basic material needs and creative potential. As such, 'full-engagement' is based on three core principles:

- that more holistic views of citizenship and engagement are required that move beyond the current focus upon workers' (i.e. employees') rights and insertion into formal employment;
- that the vast and growing proportion of informal work, such as caring, that takes place beyond employment needs to be recognised and valued; and
- that such activity is a key way in which needs are met and that many wish to engage in such work.

There is recognition, however, that if a laissez-faire approach towards informal work is continued, then current socio-economic inequalities in households' abilities to perform such work will remain.

How, therefore, can a 'full-engagement' society be created? To overcome the barriers to informal work and create sufficient activity to achieve a 'full-engagement' society, a two-pronged approach has been here advocated. First, bottom-up grass roots initiatives have been advocated to attack directly the barriers to participation in informal work and facilitate a new mutualism grounded in the development of the fourth sector of one-to-one reciprocity and second, top-down approaches to facilitate the emergence of 'worker citizens'. In Chapters 8 and 9, these bottom-up and top-down strategies were considered in turn. Common to all the initiatives considered in these chapters is their attempt to provide a means of recording, storing and rewarding participation in informal activity conducted for the good of the community.

In Chapter 8, therefore, we considered bottom-up approaches that attempt to record, store and reward informal economic activity: local exchange and trading schemes (LETS), time banks, employee mutuals and mutual aid contracts. These initiatives have been widely embraced by government as potential tools for combating social exclusion. Until now, however, they have been viewed in policy circles more as a bridge into employment than as a tool for facilitating informal economic activity (e.g. DfEE 1999, Social Exclusion Unit 2000). Here, however, we have drawn upon fresh empirical evidence to evaluate their effectiveness in both regards. This has revealed that although such initiatives can provide a springboard into the formal labour market, they are much more effective at enabling people to help themselves. If full-employment is the goal, therefore, such initiatives have only a minor role to play in tackling poverty and are of only marginal significance (Amin *et al.* 2002). If, however, the goal is to provide sufficient work (both formal and informal) and income so as to give citizens the ability to satisfy both their basic material needs and creative potential, then these initiatives have a more central place on the policy agenda. In their role of providing alternatives to capitalist production, they provide what Harvey (2000) refers to as 'spaces of hope'. As was displayed, nevertheless, such initiatives although necessary, are insufficient alone to tackle poverty and social exclusion.

Chapter 9 thus considered a variety of top-down initiatives that might make a larger impact on enabling people to help themselves and promoting a full-engagement society. In this chapter, both a Citizen's Income scheme and a particular variant, an Active Citizens' Credits scheme, were outlined that could harness informal work by recording, storing and rewarding participation in such activity. The proposal here has been that tax credits could be used not only to create an integrated tax-benefit system but also to reward people for their active citizenship.

CONCLUSIONS

In sum, our aim in this book has been to articulate and develop a new discourse on poverty. This seeks to develop informal modes of production, which, at present, the poor are less able than those attached to the formal labour market to use as an escape route from their poverty or response to their situation. Therefore, just as many have argued that intervention is required in the realm of employment to alleviate poverty, the argument of this book has been that informal modes of production also need to be assisted. To achieve this, a range of bottom-up and top-down policies to record, store and reward participation in work beyond employment have been advocated. Given that the needs and desires of many people are currently being met so inadequately by an 'employment-focused' approach, it is hoped that this tentative proposal for harnessing the informal sphere will help engender more debate on this alternative 'third way' discourse than has been so far been the case. It is, after all, not only a new model of work and welfare for the twenty-first century that is at stake but, more importantly, the livelihoods of some of the poorest people in our communities.

Bibliography

Abrahamson, P. (1992) 'Welfare pluralism: towards a new consensus for a European social policy?', in L. Hantrais, M. O'Brien and S. Mangen (eds) *The Mixed Economy of Welfare*, Loughborough: Cross-National Research Paper no. 6, European Research Centre, University of Loughborough.

ADRET (1977) *Travailler deux heures par jour*, Paris: Seuil.

Alcock, P. (1997) *Understanding Poverty*, London: Macmillan.

Allingham, M. and Sandmo, A. (1972) 'Income tax evasion: a theoretical analysis', *Journal of Public Economics* 1,3: 323–38.

Alm, J. (1991) 'A perspective on the experimental analysis of taxpayer reporting', *Accounting Review* 66,4: 577–93.

Amado, J. and Stoffaes, C. (1980) 'Vers une socio-economie duale', in A. Danzin, A. Boublil and J. Lagarde (eds) *La Société française et la technologie*, Paris: Documentation Française.

Amin, A., Cameron, A. and Hudson, R. (1999) 'Welfare as work? The potential of the UK social economy', *Environment and Planning A* 31: 2,033–51.

—— (2002) *Placing the Social Economy*, London: Routledge.

Archibugi, F. (2000) *The Associative Economy: insights beyond the welfare state and into post-capitalism*, London: Macmillan.

Atkinson, A. and Micklewright, J. (1991) 'Unemployment compensation and labour market transitions: a critical review', *Journal of Economic Literature* 29,10: 1,679–727.

Atkinson, A.B. (1998) *Poverty in Europe*, Oxford: Blackwell.

Aznar, G. (1981) *Tous a mi-temps, ou le scénario bleu*, Paris: Seuil.

Barnes, H., North, P. and Walker, P. (1996) *LETS on a Low Income*, London: New Economics Foundation.

Barr, N. (1992) 'Economic theory and the welfare state: a survey and interpretation', *Journal of Economic Literature* 30,6: 741–803.

Barthe, M.-A. (1985) 'Chomage, travail au noir et entraide familial', *Consommation* 3: 23–42.

—— (1988) *L'Economie cachée*, Paris: Syros Alternatives.

Beck, U. (2000) *The Brave New World of Work*, Cambridge: Polity.

Beneria, L. (1999) 'The enduring debate over unpaid labour', *International Labour Review* 138: 287–309.

Bennett, F. and Hirsch, D. (2001) 'Balancing support and opportunity', in F. Bennett and D. Hirsch (eds) *The Employment Tax Credit and Issues for the Future of In-Work Support*, York: York Publishing Services.

—— and Walker, R. (1998) *Working with Work: an initial assessment of welfare to work*, York: Joseph Rowntree Foundation.

Bennington, J., Baine, S. and Russell, J. (1992) 'The impact of the Single European Market on regional and local economic development and the voluntary and community sectors', in L. Hantrais, M. O'Brien and S. Mangen (eds) *The Mixed Economy of Welfare*, Loughborough: Cross-National Research Paper no. 6, European Research Centre, University of Loughborough.

Bentley, T. and Mulgan, G. (1996) *Employee Mutuals: the 21st Century trade union?*, London: Demos.

Benton, L. (1990) *Invisible Factories: the informal economy and industrial development in Spain*, New York: State University of New York Press.

Berking, H. (1999) *Sociology of Giving*, London: Sage.

Beveridge, W. (1944) *Full-employment in a Free Society*, London: George Allen and Unwin.

Bittman, M. (1995) 'The politics of the study of unpaid work', *Just Policy* 2,1: 3–10.

Blair, T. (1998) *The Third Way: new politics for the new century*, London: Fabian Society.

Booth, W. (1994) *Households*, New York: Ithaca Press.

Boswell, J. (1990) *Community and the Economy*, London: Routledge.

Boyle, D. (1999) *Funny Money: in search of alternative cash*, London: HarperCollins.

—— (2002) 'Review of Bridges into work: an evaluation of Local Exchange Trading Schemes by C.C. Williams *et al.*', *Local Economy* 17,2: 83–4.

Bradbury, B. (1989) 'Family size equivalence scales and survey evaluations of income and well-being', *Journal of Social Policy* 18,3: 383–408.

Bridges, W. (1995) *Jobshift: how to prosper in a workplace without jobs*, London: Nicholas Brealey.

Briscoe, I. (1995) *In Whose Service? Making community service work for the unemployed*, London: Demos.

Bryson, L. (1996) 'Revaluing the household economy', *Women's Studies International Forum* 19,2: 207–19.

Burns, D. and Taylor, M. (1998) *Mutual Aid and Self-Help: coping strategies for excluded communities*, Bristol: The Policy Press.

Button, K. (1984) 'Regional variations in the irregular economy: a study of possible trends', *Regional Studies* 18: 385–92.

Cabinet Office (1999) *Sharing the Nation's Prosperity: variation in economic and social conditions across the UK*, London: Cabinet Office.

—— (2000) *Sharing the Nation's Prosperity: economic, social and environmental conditions in the countryside*, London: Cabinet Office.

Cahn, E. (1994) 'Reinventing poverty law', *Yale Law Journal* 103: 2133–55.

—— (2000) *No More Throw-Away People: the co-production imperative*, Washington DC: Essential Books.

—— and Rowe, J. (1992) *Time Dollars: the new currency that enables Americans to turn their hidden resource – time – into personal security and community renewal*, Chicago, IL: Family Resource Coalition of America.

Callan, T. and Nolan, B. (1991) 'Concepts of poverty and the poverty line', *Journal of Economic Surveys* 5,3: 243–61.

——, Nolan, B. and Whelan, B.J. (1993) 'Resources, deprivation and the measurement of poverty', *Journal of Social Policy* 22,2: 141–72.

Campbell, M. (2000) 'Reconnecting the long-term unemployed to labour market opportunity: the case for a local active labour market policy', *Regional Studies* 34,7: 655–68.

Cannon, D. (1994) *Generation X and the New Work Ethic*, London: Demos.

Caplow, T. (1982) 'Christian gifts and kin networks', *American Sociological Review* 47,3: 383–92.

Cappechi, V. (1989) 'The informal economy and the development of flexible specialisation in Emilia Romagna', in A. Portes, M. Castells and L.A. Benton (eds) *The Informal Economy: studies in advanced and less developing countries*, Baltimore: Johns Hopkins University Press.

Capra, F. and Spretnak, C. (1985) *Green Politics*, London: Hutchinson.

Carruthers, B.G. and Babb, S.L. (2000) *Economy/Society: markets, meanings and social structure*, Thousand Oaks, CA: Pine Forge Press.

Castells, M. (1996) *The Rise of the Network Society*, Oxford: Blackwell.

—— and Portes, A. (1989) 'World underneath: the origins, dynamics and effects of the informal economy', in A. Portes, M. Castells and L.A. Benton (eds) *The Informal Economy: studies in advanced and less developing countries*, Baltimore: Johns Hopkins University Press.

Chabaud, D., Fougeyrollas, D. and Sonthonnax, F. (1985) *Espace et temps du travail domestique*, Paris: Méridiens.

Chadeau, A. and Fouquet, A.-M. (1981) 'Peut-on mesurer le travail domestique?', *Economie et Statistique* 136: 29–42.

Chanan, G. (1999) 'Employment and the third sector: promise and misconceptions', *Local Economy* 13,4: 361–8.

Chapman, P., Phiminster, E., Shucksmith, M., Upward, R. and Vera-Toscano, E. (1998) *Poverty and exclusion in rural Britain: the dynamics of low income and employment*, York: York Publishing Services.

Cheal, D. (1988) *The Gift Economy*, London: Verso.

Citro, C.F. and Michael, R.T. (1995) (eds) *Measuring Poverty: a new approach*, Washington DC: National Academy Press.

Cloke, P., Milbourne, P. and Thomas, C. (1994) *Lifestyles in Rural England*, London: Rural Research Report 18, Rural Development Commission.

Coleman, J. (1988) 'Social capital in the creation of human capital', *American Journal of Sociology* (Suppl.) 94: S95–120.

Community Development Foundation (1995) *Added Value and Changing Values: community involvement in urban regeneration: a 12 country study for the European Union*, Brussels: CEC DG XVI.

Conroy, P. (1996) *Equal Opportunities for All*, Brussels: European Social Policy Forum Working Paper I, DG V, European Commission.

Cook, D. (1997) *Poverty, Crime and Punishment*, London: Child Poverty Action Group.

Cornuel, D. and Duriez, B. (1985) 'Local exchange and state intervention', in N. Redclift and E. Mingione (eds) *Beyond Employment: household, gender and subsistence*, Oxford: Basil Blackwell.

Corrigan, P. (1989) 'Gender and the gift: the case of the family clothing economy', *Sociology* 23: 513–34.

Countryside Agency (2001) *The State of the Countryside 2001*, London: Countryside Agency.

Coupland, D. (1991) *Generation X: tales for an accelerated culture*, New York: St Martin's Press.

Crang, P. (1996) 'Displacement, consumption and identity', *Environment and Planning A* 28,1: 47–67.

Crewe, L. and Gregson, N. (1998) 'Tales of the unexpected: exploring car boot sales as marginal spaces of consumption', *Transactions of the Institute of British Geographers* 23,1: 39–53.

Culpitt, I. (1992) *Welfare and Citizenship*, London: Sage.

Davidson, M., Redshaw, J. and Mooney, A. (1997) *The Role of DIY in Maintaining Owner-occupied Stock*, Bristol: Policy Press.

Davies, J. (1992) *Exchange*, Milton Keynes: Open University Press.

Davies, R.B., Elias, P. and Penn, R. (1992) 'The relationship between a husband's unemployment and his wife's participation in the labour force', *Oxford Bulletin of Economics and Statistics* 54,2: 145–71.

Dawes, L. (1993) *Long-term Unemployment and Labour Market Flexibility*, Leicester: Centre for Labour Market Studies, University of Leicester.

Deakin, S. and Wilkinson, F. (1991/2) 'Social policy and economic efficiency: the deregulation of labour markets in Britain', *Critical Social Policy* 33,1: 40–51.

Dean, H. and Melrose, M. (1996) 'Unravelling citizenship: the significance of social security benefit fraud', *Critical Social Policy* 16,1: 3–31.

Deleeck, H., van Den Bosch, K. and de Lathouwer, L. (1992) (eds) *Poverty and the Adequacy of Social Security in the EC*, Aldershot: Avebury.

Delors, J. (1979) 'Le troisième secteur: le travail au-delà de l'emploi', *Autrement* 20: 147–52.

Delphy, C. (1984) *Close to Home*, London: Hutchinson.

Desai, M. (1998) *A Basic Income Proposal*, London: Social Market Foundation.

DETR (1998) *Community-based Regeneration Initiatives: a working paper*, London: DETR.

Devall, B. (1990) *Simple in Means, Rich in Ends: practising deep ecology*, London: Green Print.

—— and Sessions, G. (1985) *Deep Ecology: living as if nature mattered*, Salt Lake City: Peregrine Smith Books.

DfEE (1999) *Jobs for All: national strategy for neighbourhood renewal: PAT 1*, Nottingham: DfEE.

Dickens, R., Gregg, P. and Wadsworth, J. (2000) 'New Labour and the labour market', *Oxford Review of Economic Policy* 16,1: 95–113.

Dilnot, A. (1992) 'Social security and labour market policy', in I.E. McLaughlin (ed.) *Understanding Employment*, London: Routledge.

Dobson, R.V.G. (1993) *Bringing the Economy Home from the Market*, New York: Black Rose Books.

Donnison, D. (1998) *Policies for a Just Society*, Basingstoke: Macmillan.

Dorling, D. and Woodward, R. (1996) 'Social polarisation 1971–1991: a micro-geographical analysis of Britain', *Progress in Planning* 45,2: 67–122.

Dorsett, R. (2001) *Workless Couples: characteristics and labour market transitions*, London: Employment Service.

Douthwaite, R. (1996) *Short Circuit: strengthening local economies for security in an unstable world*, Dartington: Green Books.

DSS (1998) *New Ambitions for our Country: a new contract for welfare*, Cmnd 3805, London: The Stationery Office.

Dumontier, F. and Pan Ke Shon, J.-L. (1999) *En 13 ans, moins de temps contraints et plus de loisirs*, Paris: INSEE.

Dunford, M. (1997) 'Diversity, instability and exclusion: regional dynamics in Great Britain', in R. Lee and J. Wills (eds) *Geographies of Economies*, London: Arnold.

Eckersley, R. (1992) *Environmentalism and Political Theory: towards an ecocentric approach*, London: UCL Press.

Economist Intelligence Unit (1982) *Coping with Unemployment: the effects on the unemployed themselves*, London: Economist Intelligence Unit.

ECOTEC (1998) *Third System and Employment: evaluation inception report*, Birmingham: ECOTEC.

Eisenschitz, A. (1997) 'The view from the grassroots', in M. Pacione (ed.) *Britain's Cities: geographies of division in urban Britain*, London: Routledge.

Ekins, P. and Max-Neef, M. (1992) *Real-Life Economics: understanding wealth creation*, London: Routledge.

Elias, P. (1997) 'The effect of unemployment benefits on the labour force participation of partners', Paper presented at the 8th Annual Meeting of the European Association of Labour Economists, Aarhus, 18–20 September.

Elkin, T. and McLaren, D. (1991) *Reviving the City: towards sustainable urban development*, London: Friends of the Earth.

Elson, D. (1988) 'Market socialism or socialization of the market?', *New Left Review* 172,11: 29.

Engbersen, G., Schuytt, K., Timmer, J. and van Waarden, F. (1993) *Cultures of Unemployment: a comparative look at long-term unemployment and urban poverty*, Oxford: Westview.

Esping-Andersen, G. (1994) 'Welfare states and the economy', in N.J. Smelser and R. Swedberg (eds) *The Handbook of Economic Sociology*, Princeton, NJ: Princeton University Press.

European Commission (1996a) *For a Europe of Civic and Social Rights: report by the Comité des Sages*, Brussels: European Commission DG for Employment, Industrial Relations and Social Affairs.

—— (1996b) *Social and Economic Inclusion Through Regional Development: the community economic development priority in ESF programmes in Great Britain*, Brussels: European Commission.

—— (1997) *Towards an Urban Agenda in the European Union*, COM(97) 197, Brussels: Communication from the European Commission.

—— (1998a) *On Undeclared Work*, COM (1998) 219, Brussels: Commission of the European Communities.

—— (1998b) *The Era of Tailor-Made Jobs: second report on local development and employment initiatives*, Brussels: European Commission.

—— (1999) *Sixth Periodic Report on the Regions*, Brussels: European Commission DGXVI.

—— (2000a) *The Social Situation in the European Union 2000*, Brussels: European Commission.

—— (2000b) *Employment in Europe 2000*, Brussels: European Commission.

—— (2001a) *EU Employment and Social Policy, 1999–2001: job, cohesion, productivity*, Luxembourg: Office for Official Publications of the European Communities.

—— (2001b) *The Social Situation in the European Union 2001*, Brussels: Commission of the European Communities.

Evason, E. and Woods, R. (1995) 'Poverty, deregulation of the labour market and benefit fraud', *Social Policy and Administration* 29,1: 40–55.

Evers, A. and Wintersberger, H. (1988) (eds) *Shifts in the Welfare Mix: their impact on work, social services and welfare policies*, Vienna: European Centre for Social Welfare Training and Research.

Fainstein, N. (1996) 'A note on interpreting American poverty', in E. Mingione (ed.) *Urban Poverty and the Underclass*, Oxford: Basil Blackwell.

Falkinger, J. (1988) 'Tax evasion and equity: a theoretical analysis', *Public Finance* 43: 388–95.

Feige, E.L. (1979) 'How big is the irregular economy?', *Challenge* November/December: 5–13.

—— (1990) 'Defining and estimating underground and informal economies', *World Development* 18: 989–1,002.

Findlay, A., Short, D., Stockdale, A. (1999) *Migration impacts in rural England*, Cheltenham: CAX 19, Countryside Agency.

Fitzpatrick, T. (1999) *Freedom and Security: an introduction to the basic income debate*, London: Macmillan.

Fodor, E. (1999) *Better not Bigger: how to take control of urban growth and improve your community*, Stony Creek, NJ: New Society.

Fordham, G. (1995) *Made to Last: creating sustainable neighbourhood and estate regeneration*, York: Joseph Rowntree Foundation.

Fortin, B., Garneau, G., Lacroix, G., Lemieux, T. and Montmarquette, C. (1996) *L'Economie souterraine au Quebec: mythes et réalités*, Laval: Presses de l'Université Laval.

Foudi, R., Stankiewicz, F. and Vanecloo, N. (1982) 'Chomeurs et économie informelle', *Cahiers de l'observation du changement social et culturel no. 17*, Paris: CNRS.

Franks, S. (2000) *Having None of It: women, men and the future of work*, London: Granta.

Friedmann, Y. (1982) 'L'art de la survie', *Autogestions* 8/9: 89–95.

Gardiner, J. (1997) *Gender, Care and Economics*, Basingstoke: Macmillan.

Gass, R. (1996) 'The next stage of structural change: towards a decentralised economy and an active society', in OECD (ed.) *Reconciling Economy and Society: towards a plural economy*, Paris: OECD.

Gershuny, J. (1985) 'Economic development and change in the mode of provision of services', in N. Redclift and E. Mingione (eds) *Beyond Employment: household, gender and subsistence*, Oxford: Basil Blackwell.

—— (2000) *Changing Times: work and leisure in post-industrial society*, Oxford: Oxford University Press.

—— and Jones, S. (1987) 'The changing work/leisure balance in Britain 1961–84', *Sociological Review Monograph* 33: 9–50.

—— and Miles, I. (1983) *The New Service Economy: the transformation of employment in industrial societies*, London: Frances Pinter.

——, Godwin, M. and Jones, S. (1994) 'The domestic labour revolution: a process of lagged adaptation', in M. Anderson, F. Bechhofer and J. Gershuny (eds) *The Social and Political Economy of the Household*, Oxford: Oxford University Press.

Gertler, M. (1997) 'The invention of regional culture', in R. Lee and J. Wills (eds) *Geographies of Economies*, London: Arnold.

Gibson-Graham, J.K. (1996) *The End of Capitalism (as we knew it)? A feminist critique of political economy*, London: Blackwell.

Giddens, A. (1998) *The Third Way: the renewal of social democracy*, Cambridge: Polity Press.

—— (2000), *The Third Way and its Critics*, Cambridge: Polity.

—— (2002) *Where Now for New Labour?*, Cambridge: Polity.

Gilder, G. (1981) *Wealth and Poverty*, New York: Basic Books.

Glatzer, W. and Berger, R. (1988) 'Household composition, social networks and household production in Germany', in R. Pahl (ed.) *On Work: historical, comparative and theoretical approaches*, Oxford: Basil Blackwell.

Goldschmidt-Clermond, L. (1982) *Unpaid Work in the Household: a review of economic evaluation methods*, Geneva: ILO.

Goldsmith, E., Khor, M., Norberg-Hodge, H. and Shiva, V. (1995) (eds) *The Future of Progress: reflections on environment and development*, Dartington: Green Books.

Goodin, R. (1992) *Green Political Theory*, Cambridge: Polity.

Gordon, D., Adelam, L., Ashworth, K., Bradshaw, L., Levitas, L., Middleton, S., Pantazis, C., Patisos, D., Payne, S., Townsend, P. and Williams, J. (2000) *Poverty and Social Exclusion in Britain*, York: Joseph Rowntree Foundation.

Gorz, A. (1985) *Paths to Paradise*, London: Pluto.

—— (1999) *Reclaiming Work: beyond the wage-based society*, Cambridge: Polity.

Gough, I. (2000) *Global Capital, Human Needs and Social Policies*, Basingstoke: Palgrave.

Government Statistical Service (1998) *Harmonised Concepts and Questions for Government Social Surveys*, London: Government Statistical Service.

Grabiner, Lord (2000) *The Informal Economy*, London: HM Treasury.

Granovetter, M. (1973) 'The strength of weak ties', *American Journal of Sociology* 78: 1,360–80.

Gray, J. (1997) *Enlightenment's Wake*, London: Routledge.

Green, A.E. and Owen, D. (1998) *Where are the Jobless? Changing unemployment and non-employment in cities and regions*, York: The Policy Press.

Green, D. (1993) *Reinventing Civil Society: the rediscovery of welfare without politics*, London: Institute for Economic Affairs.

Greffe, X. (1981) 'L'économie non-officielle', *Consommation* 3: 5–16.

Gregg, P. and Wadsworth, J. (1996) *It Takes Two: employment polarisation in the OECD*, London: Discussion Paper no. 304, Centre for Economic Performance, London School of Economics.

——, Hansen, K. and Wadsworth, J. (1999) 'The rise of the workless household', in P. Gregg and J. Wadsworth (eds) *The State of Working Britain*, Manchester: Manchester University Press.

Gregory, A. and Windebank, J. (2000) *Women and Work in France and Britain: theory, practice and policy*, Basingstoke: Macmillan.

Gregson, N. and Crewe, L. (2002) *Second-hand Worlds*, London: Routledge.

Gudeman, S. (2001) *The Anthropology of Economy*, Oxford: Blackwell.

Gutmann, P.M. (1978) 'Are the unemployed, unemployed?', *Financial Analysts Journal* 35: 26–7.

Habermas, J. (1975) *Legitimation crisis*, London: Heinemann.

Haicault, M. (1984) 'La gestion ordinaire de la vie en deux', *Sociologie du Travail* 3: 268–77.

Hakim, C. (2000) *Work-Lifestyle Choices in the 21st century*, Oxford: Oxford University Press.

Hall, P.A. (1997) 'Social capital: a fragile asset', in I. Christie and S. Perry (eds) *The Wealth and Poverty of Networks*, London: Demos.

Hall, S. (1999) 'Ministers give carers a £140m break', *Guardian* 9 February: 9.

Hallerod, B. (1995) 'Perceptions of poverty in Sweden', *Scandanavian Journal of Social Welfare* 4,3: 174–89.

Hamnett, C. (2001) 'The emperor's new theoretical clothes, or geography without origami', in G. Philo and D. Miller (eds) *Market Killing: what the free market does and what social scientists can do about it*, London: Longman.

Harding, P. and Jenkins, R. (1989) *The Myth of the Hidden Economy: towards a new understanding of informal economic activity*, Milton Keynes: Open University Press.

Hargreaves, I. (1998) 'A step beyond Morris dancing: the third sector revival', in I. Hargreaves and I. Christie (eds) *Tomorrow's Politics: the Third Way and beyond*, London: Demos.

Harkness, S. (1994) 'Female employment and changes in the share of women's earnings in total family income in Great Britain', in S. Hardy, G. Lloyd and I. Cundell (eds)

Tackling Unemployment and Social Exclusion: problems for regions, solutions for people, London: Regional Studies Association.

Harvey, D. (1982) *The Limits to Capital*, Oxford: Blackwell.

——— (1989) *The Condition of Postmodernity: an enquiry into the origins of cultural change*, Oxford: Blackwell.

——— (2000) *Spaces of Hope*, Edinburgh: Edinburgh University Press.

Hasseldine, J. and Zhuhong, L. (1999) 'More tax evasion research required in the new millennium', *Crime, Law and Social Change* 31: 91–104.

Haughton, G. (1998) 'Principles and practice of community economic development', *Regional Studies* 32,9: 872–8.

———, Johnson, S., Murphy, L. and Thomas, K. (1993) *Local Geographies of Unemployment: long-term unemployment in areas of local deprivation*, Aldershot: Avebury.

Hedges, A. (1999) *Living in the Countryside: the needs and aspirations of rural populations*, London: Countryside Agency.

Heinze, R.G. and Olk, T. (1982) 'Development of the informal economy: a strategy for resolving the crisis of the welfare state', *Futures* 4: 189–204.

Henderson, H. (1999) *Beyond Globalisation: shaping a sustainable global economy*, London: Kumarian Press.

Hetherington, K. (1998) *Expressions of Identity: space, performance, politics*. London: Sage.

Hewitt, P. (1994) 'Full-employment for men and women', paper presented at TUC Conference on *Looking forward and full-employment*, Brighton, July.

Hills, J. (1998) *Thatcherism, New Labour and the Welfare State*, London: Centre for Analysis of Social Exclusion Paper 13, London School of Economics.

Himmelweit, S. (2000) (ed.) *Inside the Household: from labour to care*, Basingstoke: Macmillan.

Hirsch, D. (1999) *Welfare beyond Work: active participation in a new welfare state*, York: York Publishing Services.

HM Treasury (1997) *Employment Opportunity in a Changing Labour Market*, London: HM Treasury.

——— (1998) *The Modernisation of Britain's Tax and Benefit System: the Working Families Tax Credit and work incentives*, London: HM Treasury.

Hochschild, A. (1989) *The Second Shift: working parents and the revolution at home*, New York: Viking Press.

Home Office (1999) *Community Self-help – Policy Action Team no. 9*, London: Home Office.

Hoogendijk, W. (1993) *The Economic Revolution: towards a sustainable future by freeing the economy from money-making*, Utrecht: International Books.

Howe, L. (1988) 'Unemployment, doing the double and local labour markets in Belfast', in C. Cartin and T. Wilson (eds) *Ireland from Below: social change and local communities in modern Ireland*, Dublin: Gill and Macmillan.

——— (1990) *Being Unemployed in Northern Ireland: an ethnographic study*, Cambridge: Cambridge University Press.

ILO (1996) *World Employment 1996/97: national policies in a global context*, Geneva: International Labour Office.

——— (1997) *World Employment 1997–98*, Geneva: International Labour Office.

Ironmonger, D. (1996) 'Counting outputs, capital inputs and caring labor: estimating gross household product', *Feminist Economics* 2,3: 37–64.

Isachsen, A.J., Klovland, J.T. and Strom, S. (1982) 'The hidden economy in Norway', in V. Tanzi (ed.) *The Underground Economy in the United States and Abroad*, Lexington, KY: D.C. Heath.

James, S. (1994) 'Women's unwaged work: the heart of the informal sector', in M. Evans (ed.) *The Woman Question*, London: Sage.

Jenkins, S.P. and O'Leary, N.C. (1996) 'Household income plus household production and the distribution of extended income in the UK', *Review of Income and Wealth* 42: 26–41.

Jenkins, S.P. and O'Leary, N.C. (1997) 'Gender differentials in domestic work, market work and total work time: UK time budget survey evidence for 1974/5 and 1987', *Scottish Journal of Political Economy* 44,2: 153–64.

Jensen, L., Cornwell, G.T. and Findeis, J.L. (1995) 'Informal work in non-metropolitan Pennsylvania', *Rural Sociology* 60: 91–107.

Jordan, B. (1998) *The New Politics of Welfare: social justice in a global context*, London: Sage.

—— and Jordan, C. (2000) *Social Work and the Third Way: tough love as social policy*, London: Sage.

——, James, S., Kay, H. and Redley, M. (1992) *Trapped in Poverty*, London: Routledge.

——, Agulnik, P., Burbridge, D. and Duffin, S. (2000) *Stumbling Towards Basic Income: the prospects for tax-benefit integration*, London: Citizen's Income Study Centre.

Juster, T.F. and Stafford, F.P. (1991) 'The allocation of time: empirical findings, behavioural models and problems of measurement', *Journal of Economic Literature* 29,2: 471–522.

Keck, M. and Sikkunk, K. (1998) *Activist Beyond Borders*, Ithaca, NY: Princeton University Press.

Kempson, E. (1996) *Life on a Low Income*, York: York Publishing Services.

Kesteloot, C. and Meert, H. (1999) 'Informal spaces: the geography of informal economic activities in Brussels', *International Journal of Urban and Regional Research* 23: 232–51.

King, D. and Wickham-Jones, M. (1999) 'Bridging the Atlantic: the Democratic (Party) origins of welfare to work', in M. Powell (ed.) *New Labour, New Welfare State? The 'third way' in British social policy*, Bristol: The Policy Press.

Komter, A.E. (1996) 'Reciprocity as a principle of exclusion: gift giving in the Netherlands', *Sociology* 30: 299–316.

Koopmans, C.C. (1989) *Informele Arbeid: vraag. aanbod, participanten, prijzen*, Amsterdam: Proefschrift Universitiet van Amsterdam.

Lagos, R.A. (1995) 'Formalising the informal sector: barriers and costs', *Development and Change* 26: 110–31.

Lalonde, B. and Simmonet, D. (1978) *Quand vous voudrez*, Paris: Pauvert.

Laville, J.-L. (1995) 'La crise de la condition salariale: emploi, activité et nouvelle question sociale', *Esprit* 12: 32–54.

—— (1996) 'Economy and solidarity: exploring the issues', in OECD (ed.) *Reconciling Economy and Society: towards a plural economy*, Paris: OECD.

Leadbeater, C. (1999) *Living on Thin Air: the new economy*, Harmondsworth: Penguin.

—— and Martin, S. (1998) *The Employee Mutual: combining flexibility with security in the new world of work*, London: Demos.

Leather, P., Littlewood, A. and Munro, M. (1998) *Make Do and Mend: explaining homeowners' approaches to repair and maintenance*, Bristol: The Policy Press.

Lebreton, P. (1978) *L'ex-Croissance: les chemins de l'écologisme*, Paris: Denoel.

Lee, R. (1996) 'Moral money? Making local economic geographies: LETS in Kent, southeast England', *Environment and Planning A* 27,8: 1,377–94.

—— (1997) 'Economic geographies: representations and interpretations', in R. Lee and J. Wills (eds) *Geographies of Economies*, London: Edward Arnold.

—— (1999) 'Production', in P. Cloke, P. Crang and M. Goodwin (eds) *Introducing Human Geographies*, London: Arnold.

—— (2000a) 'Shelter from the storm? Geographies of regard in the worlds of horticultural consumption and production', *Geoforum* 31: 137–57.

—— (2000b) 'Economic geography', in R.J. Johnston, D. Gregory, G. Pratt and M. Watts (eds) *The Dictionary of Human Geography*, Oxford: Blackwell.

Lemieux, T., Fortin, B. and Frechette, P. (1994) 'The effect of taxes on labor supply in the underground economy', *American Economic Review* 84: 231–54.

Leonard, M. (1994) *Informal Economic Activity in Belfast*, Aldershot: Avebury.

—— (1998) *Invisible Work, Invisible Workers: the informal economy in Europe and the US*, Basingstoke: Macmillan.

Levitas, R. (1998) *The Inclusive Society? social exclusion and New Labour*, Basingstoke: Macmillan.

Liebman, J. (1998) *Lessons about Tax-Benefit Integration from the US Earned Income Tax Credit Experience*, York: York Publishing Services.

Lindbeck, A. (1981) *Work Disincentives in the Welfare State*, Stockholm: Institute for International Economic Studies, University of Stockholm.

Lipietz, A. (1992) *Towards a New Economic Order: post-Fordism, ecology and democracy*, Cambridge: Polity.

—— (1995) *Green Hopes: the future of political ecology*, Cambridge: Polity.

Lister, R. (1997) *Citizenship: feminist perspectives*, Basingstoke: Macmillan.

—— (1998) 'From equality to social inclusion: New Labour and the welfare state', *Critical Social Policy* 18,2: 215–25.

—— (2000) 'Strategies for social inclusion: promoting social cohesion or social justice?', in P. Askonas and A. Stewart (eds) *Social Inclusion: possibilities and tensions*, Basingstoke: Macmillan.

Lobo, F.M. (1990a) 'Irregular work in Spain', in *Underground Economy and Irregular Forms of Employment, Final Synthesis Report*, Brussels: Office for Official Publications of the European Communities.

—— (1990b) 'Irregular work in Portugal', in *Underground Economy and Irregular Forms of Employment, Final Synthesis Report*, Brussels: Office for Official Publications of the European Communities.

Lodemel, I. and Trickey, H. (2001) *An Offer You Can't Refuse: workfare in international perspective*, Bristol: The Policy Press.

Lorendahl, B. (1997) 'Integrating public and co-operative/third sector: towards a new Swedish model', *Annals of Public and Co-operative Economics* 68,3: 379–96.

Lozano, B. (1989) *The Invisible Workforce: transforming American business with outside and home-based workers*, New York: The Free Press.

Luxton, M. (1997) 'The UN, women and household labour: measuring and valuing unpaid work', *Women's Studies International Forum* 20: 431–9.

MacDonald, R. (1994) 'Fiddly jobs, undeclared working and the something for nothing society', *Work, Employment and Society* 8: 507–30.

Macfarlane, R. (1996) *Unshackling the Poor: a complementary approach to local economic development*, York: Joseph Rowntree Foundation.

Maffesoli, M. (1996) *The Time of the Tribes: the decline of individualism in mass society*, London: Sage.

Maldonado, C. (1995) 'The informal sector: legalization or laissez-faire?', *International Labour Review* 134: 705–28.

Mander, J. and Goldsmith, E. (1996) (eds) *The Case Against the Global Economy: and for a turn toward the local*, San Francisco: Sierra Club.

Markusen, A. (1999) 'Fuzzy concepts, scanty evidence and policy distance: the case for rigour and policy relevance in critical regional studies', *Regional Studies* 33: 869–86.

Martin, R. (2001) 'Geography and public policy: the case of the missing agenda', *Progress in Human Geography* 25,2: 189–210.

Mattera, P. (1980) 'Small is not beautiful: decentralised production and the underground economy in Italy', *Radical America* 14: 67–76.

Matthews, K. (1983) 'National income and the black economy', *Journal of Economic Affairs* 3: 261–7.

Mauss, M. (1966) *The Gift*, London: Cohen and West.

Mauthner, N., McKee, L. and Strell, M. (2001) *Work and Family Life in Rural Communities*, York: York Publishing Services.

Mayer, M. and Katz, S. (1985) 'Gimme shelter: self-help housing struggles within and against the state in New York City and West Berlin', *International Journal of Urban and Regional Research* 9: 123–56.

Mayo, E. (1996) 'Dreaming of work', in P. Meadows (ed.) *Work Out or Work In? Contributions to the debate on the future of work*, York: Joseph Rowntree Foundation.

McBurney, S. (1990) *Ecology into Economics Won't Go: or life is not a concept*, Dartington: Green Books.

McCormick, J. (1994) *Citizens' Service*, London: Institute for Public Policy Research.

—— and Oppenheim, C. (1998) (eds) *Welfare in Working Order*, London: Institute for Public Policy Research.

McDowell, L. (1991) 'Life without father and Ford: the new gender order of post-Fordism', *Transactions of the Institute of British Geographers* 16,4: 400–19.

—— (2001) 'Father and Ford revisited: gender, class and employment change in the new millennium', *Transactions of the Institute of British Geographers* NS 26,4: 448–64.

McGlone, F., Park, A. and Smith, K. (1998) *Families and Kinship*, London: Family Policy Studies Centre.

McKay, A. and Vanavery, J. (2000) 'Gender, family and income maintenance: a feminist case for citizen's basic income', *Social Politics* 7,2: 266–84.

Mckay, G. (1998) (ed.) *DiY Culture*, London: Verso.

McLaughlin, E. (1994) *Flexibility in Work and Benefits*, London: Institute of Public Policy Research.

Meadows, P. (1997) *The Integration of Taxes and Benefits for Working Families with Children: issues raised to date*, York: York Publishing Services.

Meehan, E. (1993) *Citizenship and the European Community*, London: Sage.

Meert, H., Mestiaen, P. and Kesteloot, C. (1997) 'The geography of survival: household strategies in urban settings', *Tijdschrift voor Economische en Sociale Geografie* 88,2: 169–81.

Millar, J. and Hole, D. (1998) *Integrated Family Benefits in Australia and Options for the UK Tax Return System*, York: York Publishing Services.

Milliron, V. and Toy, D. (1988) 'Tax compliance: an investigation of key features', *The Journal of the American Taxation Association* 9: 84–104.

Minc, A. (1980) 'Le chomage et l'économie souterraine', *Le Debat* 2: 3–14.

—— (1982) *L'Après-Crise à Commencé*, Paris: Gallimard.

Mingione, E. (1991) *Fragmented Societies: a sociology of economic life beyond the market paradigm*, Oxford: Basil Blackwell.

—— and Morlicchio, E. (1993) 'New forms of urban poverty in Italy: risk path models in the North and South', *International Journal of Urban and Regional Research* 17: 413–27.

Mogensen, G.V. (1985) *Sort Arbejde i Danmark*, Copenhagen: Institut for National-okonomi.

Morehouse, W. (1997) (ed.) *Building Sustainable Communities: tools and concepts for self-reliant economic change*, Charlbury: Jon Carpenter.

Morris, L. (1987) 'Local social polarisation: a case study of Hartlepool', *International Journal of Urban and Regional Research* 11,3: 351–62.

—— (1994) 'Informal aspects of social divisions', *International Journal of Urban and Regional Research* 18: 112–26.

—— (1995) *Social Divisions: economic decline and social structural change*, London: UCL Press.

Murgatroyd, L. and Neuburger, H. (1997) 'A household satellite account for the UK', *Economic Trends* 527: 63–71.

Murray, C. (1984) *Losing Ground: American social policy, 1950–1980*, New York: Basic Books.

Myles, J. (1996) 'When markets fail: social welfare in Canada and the US', in G. Esping-Anderson (ed.) *Welfare States in Transition: national adaptations in global economies*, London: Sage.

Naess, A. (1986) 'The deep ecology movement: some philosophical aspects', *Philosophical Inquiry* III, 1/2: 10–31.

—— (1989) *Ecology, Community and Lifestyle: outline of an ecosophy*, Cambridge: Cambridge University Press.

Nelson, M.K. and Smith, J. (1999) *Working Hard and Making Do: surviving in small town America*, Los Angeles: University of California Press.

New Earnings Survey (1998) London: ONS.

Nicaise, I. (1996) *Which Partnerships for Employment? Social partners, NGOs and public authorities*, Brussels: European Social Policy Forum Working Paper II, DG V, European Commission.

North, P. (1996) 'LETS: a tool for empowerment in the inner city?', *Local Economy* 11: 284–93.

—— (1998) 'Exploring the politics of social movements through "sociological intervention": a case study of Local Exchange Trading Schemes', *Sociological Review* 46: 564–82.

—— (1999) 'Explorations in heterotopia: LETS and the micropolitics of money and livelihood', *Environment and Planning D: Society and Space* 17: 69–86.

O'Connor, J. (1973) *The Fiscal Crisis of the State*, London: St Martin's Press.

OECD (1993) *Employment Outlook*, Paris: OECD.

—— (1994) *Jobs Study: Part 2*, Paris: OECD.

—— (1996) (ed.) *Reconciling Economy and Society: towards a plural economy*, Paris: OECD.

—— (1999) *The Local Dimension of Welfare-to-Work: an international survey*, Paris: OECD.

—— (2000) 'Rewarding work', *Employment Outlook*, June, 7–10.

Offe, C. (1995) 'Freiwillig auf die Anteilnahme am Arbeitmarkt verzichten', *Frankfurter Rundschau*, 19 July: pp. 5–6.

Okun, A.M. (1975) *Equality and Efficiency: the big trade-off*, Washington DC: Brookings Institute.

Oppenheim, C. (1998) 'Welfare to work: taxes and benefits', in J. McCormick and C. Oppenheim (eds) *Welfare in Working Order*, London: Institute for Public Policy Research.

O'Riordan, T. (1996) 'Environmentalism on the move', in I. Douglas, R. Huggett and M. Robinson (eds) *Companion Encyclopaedia of Geography*, London: Routledge.

Oxley, H. (1999) 'Income dynamics: inter-generational evidence', in Centre for Analysis of Social Exclusion (ed.) *Persistent Poverty and Lifetime Inequality: the evidence*, London: CASE Report 5, London School of Economics.

Pacione, M. (1997) 'Local Exchange Trading Systems as a response to the globalisation of capitalism', *Urban Studies* 34: 1,179–99.

Pahl, R.E. (1984) *Divisions of Labour*, Oxford: Blackwell.

—— (1995) 'Finding time to live', *Demos* 5: 12–13.

Parker, H. and Sutherland, H. (1998) 'How to get rid of the poverty trap: basic income plus national wage', *Citizen's Income Bulletin* 25: 11–14.

Paugam, S. and Russell, H. (2000) 'The effects of employment precarity and unemployment on social isolation', in D. Gallie and S. Paugam (eds) *Welfare Regimes and the Experience of Unemployment in Europe*, Oxford: Oxford University Press.

Peck, J. (1996) *Work-Place: the social regulation of labour markets*, London: Guildford Press.

—— (1998) 'Editorial: grey geography', *Transactions of the Institute of British Geographers* NS 24: 131–5.

—— (1999) 'New labourers? Making a new deal for the "workless class"', *Environment and Planning C: Government and Policy* 17: 345–72.

—— (2001) *Workfare States*, London: Guildford Press.

Perri 6 (1997) *The Power to Bind and Loose: tackling network poverty*, London: Demos.

Perrons, D. (1995) 'Gender inequalities in regional development', *Regional Studies* 29,5: 465–76.

Pestoff, V.A. (1996) 'Work environment and social enterprises in Sweden', paper presented to the *European Conference on Labour Markets, Unemployment and Co-ops*, Budapest, 27–8 October.

Peterson, H.G. (1982) 'Size of the public sector, economic growth and the informal economy: development trends in the Federal Republic of Germany', *Review of Income and Wealth* 28: 191–215.

Pinch, S. (1993) 'Social polarisation: a comparison of evidence from Britain and the United States', *Environment and Planning A* 25: 779–95.

Polanyi, K. (1944) *The Great Transformation*, Boston: Beacon Press.

Porritt, J. (1996) 'Local jobs depend on local initiative', *Finance North*, September/ October: 88.

Portes, A. (1994) 'The informal economy and its paradoxes', in N.J. Smelser and R. Swedberg (eds) *The Handbook of Economic Sociology*, Princeton: Princeton University Press.

Powell, M. (1999) (ed.) *New Labour, New Welfare State? The 'third way' in British social policy*, Bristol: The Policy Press.

Putnam, R. (1993) *Making Democracy Work: civic traditions in modern Italy*, Princeton: Princeton University Press.

—— (1995a) 'Tuning in, tuning out: the strange disappearance of social capital in America', *Political Science and Politics* 28: 664–83.

—— (1995b) 'Bowling alone: America's declining social capital', *Journal of Democracy* 6: 65–78.

—— (2000) *Bowling Alone: the collapse and revival of American community*, London: Simon and Schuster.

Putterman, L. (1990) *Division of Labour and Welfare: an introduction to economic systems*, Oxford: Oxford University Press.

Reid, M. (1934) *Economics of Household Production*, New York: John Wiley.

Renooy, P. (1990) *The Informal Economy: meaning, measurement and social significance*, Amsterdam: Netherlands Geographical Studies no. 115.

Reward Group (1999) *Cost of Living Report – Town Comparison: Southampton and Sheffield*, London: Reward Group.

Rifkin, J. (1995) *The End of Work*, New York: G.P. Putnam's.

Robertson, J. (1981) 'The future of work: some thoughts about the roles of men and women in the transition to a SHE future', *Women's Studies International Quarterly* 4: 83–94.

—— (1985) *Future Work: jobs, self-employment and leisure after the industrial age*, Aldershot: Gower/Temple Smith.

—— (1991) *Future Wealth: a new economics for the 21st century*, London: Cassells.

—— (1998) *Beyond the Dependency Culture: people, power and responsibility*, London: Adamantine.

Robinson, J. and Godbey, G. (1997) *Time for Life: the surprising ways Americans use their time*, Pennsylvania: Pennsylvania State University Press.

Robinson, P. (1998) 'Employment and social inclusion', in C. Oppenheim (ed.) *An Inclusive Society: strategies for tackling poverty*, London: Institute for Public Policy Research.

Robson, B.T. (1988) *Those Inner Cities: reconciling the social and economic aims of urban policy*, Oxford: Clarendon.

Roche, M. (2000) *Social Exclusion and the Development of European Citizenship*, Brussels: European Commission DG12.

Rosanvallon, P. (1980) 'Le developpement de l'économie souterraine et l'avenir des société industrielles', *Le Debat* 2: 8–23.

Roseland, M. (1998) (ed.) *Towards Sustainable Communities: resources for citizens and their governments*, Stony Creek, CT: New Society Publishers.

Rostow, W.J. (1960) *The Stages of Economic Growth: a non-communist manifesto*, Cambridge: Cambridge University Press.

Roustang, G. (1987) *L'Emploi: un choix de société*, Paris: Syros.

Rowlingson, K., Whyley, C., Newburn, T. and Berthoud, R. (1997) *Social Security Fraud*, London: DSS Research Report no. 64, HMSO.

Roy, C. (1991) 'Les emplois du temps dans quelques pays occidentaux', *Donnes Sociales* 2: 223–5.

Sachs, I. (1984) *Development and Planning*, Cambridge: Cambridge University Press.

Sassen, S. (1989) 'New York city's informal economy', in A. Portes, M. Castells and L.A. Benton (eds) *The Informal Economy: studies in advanced and less developing countries*, Baltimore: Johns Hopkins University Press.

Sauvy, A. (1984) *Le Travail noir et l'économie de demain*, Paris: Calmann-Levy.

Sayer, A. (1997) 'The dialectic of culture and economy', in R. Lee and J. Wills (eds) *Geographies of Economies*, London: Arnold.

—— (2001) 'For a critical cultural economy', *Antipode* 33,4: 687–708.

Schor, J. (1991) *The Overworked American: the unexpected decline of leisure*, New York: Basic Books.

Scott, A.J. (2001) 'Capitalism, cities and the production of symbolic forms', *Transactions of the Institute of British Geographers* NS 26: 11–23.

Sen, A. (1998) *Inequality Re-examined*, Oxford: Clarendon.

Shucksmith (2000) *Exclusive Countryside? Social inclusion and regeneration in rural areas*, York: Joseph Rowntree Foundation.

Simey, M. (1996) 'The end of work', *Local Work* 71: 1–6.

Simon, C.P. and Witte, A.D. (1982) *Beating the System: the underground economy*, Boston, MA: Auburn House.

Skolimowski, H. (1981) *Eco-Philosophy: designing new tactics for living*, London: Marion Boyars.

Smiles, S. (1996) *Self-Help: with illustrations of conduct and perseverance*, London: Institute of Economic Affairs.

Smith, D.H. (2000) *Grassroots Associations*, London: Sage.

Smith, S. (1986) *Britain's Shadow Economy*, Oxford: Clarendon.

Social Exclusion Unit (1998) *Bringing Britain Together: a national strategy for neighbourhood renewal*, Cm 4045, London: HMSO.

—— (2000) *National Strategy for Neighbourhood Renewal: a framework for consultation*, London: Social Exclusion Unit.

Soiffer, S.S. and Herrmann, G.M. (1987) 'Visions of power: ideology and practice in the American garage sale', *Sociological Review* 35: 48–83.

Soto, de, H. (1989) *The Other Path*, London: IB Taurus.

Stoleru, L. (1982) *La France à deux vitesses*, Paris: Flammarion.

Stratford, N. and Christie, I. (2000) 'Town and country life', in R. Jowell, J. Curtice, A. Park, K. Thomson, L. Jarvis, C. Bromley and N. Stratford (eds) *British Social Attitudes, The 17th Report: focusing on diversity*, London: Sage.

Sue, R. (1995) *Temps et Ordre Social*, Paris: PUF.

Sundbo, J. (1997) 'The creation of service markets to solve political-sociological problems: the Danish Home Service', *Service Industries Journal* 17,4: 580–602.

Sunley, P. and Martin, R. (2001) 'The geographies of the national minimum wage', *Environment and Planning A*32,10: 1,735–58.

Thomas, J.J. (1992) *Informal Economic Activity*, Hemel Hempstead: Harvester Wheatsheaf.

Thomas, K. and Smith, K. (1995) 'Results of the 1993 Census of Employment', *Employment Gazette* 103: 369–84.

Thompson, E. (1991) *Customs in Common*, London: Penguin.

Thrift, N. (2000) 'Commodities', in R.J. Johnston, D. Gregory, G. Pratt and M. Watts (eds) *The Dictionary of Human Geography*, Oxford: Blackwell.

—— and Olds, K. (1996) 'Refiguring the economic in economic geography', *Progress in Human Geography* 20: 311–17.

Tievant, S. (1982) 'Vivre autrement: échanges et sociabilité en ville nouvelle', Paris: *Cahiers de l'OCS* 6, CNRS.

Townsend, A.R. (1997) *Making a Living in Europe: human geographies of economic change*, London: Routledge.

Trainer, T. (1996) *Towards a Sustainable Economy: the need for fundamental change*, Oxford: Jon Carpenter.

Turok, I. and Edge, N. (1999) *The Jobs Gap in Britain's Cities: employment loss and labour market consequences*, Bristol: The Policy Press.

UN (1995) *The Copenhagen Declaration and Programme of Action: world summit for social development 6–12 March 1995*, New York: United Nations Department of Publications.

Unger, R. (1987) *Social Theory: its situation and its task*, Cambridge: Cambridge University Press.

Urry, J. (2000) *Sociology Beyond Societies: mobilities for the twenty-first century*, London: Routledge.

Van Berkel, R. (2000) *Inclusion Through Participation*, Brussels: European Commission DG12.

Van Eck, R. and Kazemier, B. (1985) *Zwarte Inkomsten Uit Arbeid: resultaten van in 1983 gehouden experimentele enquetes*, The Hague: CBS-Statistische Katernen nr. 3, Central Bureau of Statistics.

Van Geuns, R., Mevissen, J. and Renooy, P.H. (1987) 'The spatial and sectoral diversity of the informal economy', *Tijdschrift voor Economische en Sociale Geografie* 78: 389–98.

Van Parijs, P. (1995) *Real Freedom for All: what (if anything) is wrong with capitalism?*, Oxford: Oxford University Press.

—— (2000a) *Basic Income: guaranteed income for the XXIst century?*, Barcelona: Fundació Rafael Campalans.

—— (2000b) 'Basic income and the two dilemmas of the welfare state', in C. Pierson and F.G. Castles (eds) *The Welfare State: a reader*, Cambridge: Polity.

Van Trier, W. (1995) 'Every one a king', Ph.D. dissertation, Leuven: University of Leuven, Department of Sociology.

Veit-Wilson, J.H. (1987) 'Consensual approaches to poverty lines and social security', *Journal of Social Policy* 16,2: 183–211.

Verschave, F.-X. (1996) 'The house that Braudel built: rethinking the architecture of society', in OECD (ed.) *Reconciling Economy and Society: towards a plural economy*, Paris: OECD.

Walby, S. (1997) *Gender Transformations*, London: Routledge.

Walker, R. (1987) 'Consensual approaches to the definition of poverty: towards an alternative methodology', *Journal of Social Policy* 16,2: 213–26.

Warburton, D. (1998) *Community and Sustainable Development: participation in the future*, London: Earthscan.

Ward, L. (1999) 'Daddy's home', *Guardian* 16 June: 6.

Warde, A. (1990) 'Household work strategies and forms of labour: conceptual and empirical issues', *Work, Employment and Society* 4,4: 495–515.

Warren, M.R. (1994) 'Exploitation or co-operation? The political basis of regional variation in the Italian informal economy', *Politics and Society* 22: 89–115.

Watts, M. (1999) 'Commodities', in P. Cloke, P. Crang and M. Goodwin (eds) *Introducing Human Geographies*, London: Arnold.

Weigel, R., Hessing, D. and Elffers, H. (1987) 'Tax evasion research: a critical appraisal and theoretical model', *Journal of Economic Psychology* 8: 215–35.

Westerdahl, S. and Westlund, H. (1998) 'Third sector and new jobs: a summary of twenty case studies in European regions', *Annals of Public and Co-operative Economics* 69: 193–218.

Williams, C.C. (1996a) 'Local Exchange and Trading Systems (LETS): a new form of work and credit for the poor and unemployed', *Environment and Planning A* 28,8: 1,395–415.

—— (1996b) 'The new barter economy: an appraisal of Local Exchange and Trading Systems (LETS)', *Journal of Public Policy* 16,1: 55–71.

—— (1996c) 'Informal sector responses to unemployment: an evaluation of the potential of Local Exchange and Trading Systems (LETS)', *Work, Employment and Society* 10,2: 341–59.

—— (2000) 'Local economic development', in M. Chapman and P. Allmendinger (eds) *Planning Beyond 2000*, London: Wiley.

—— (2001a) 'Does work pay? Spatial variations in the benefits of employment and coping abilities of the unemployed', *Geoforum* 32,2: 199–214.

—— (2001b) 'Tackling the participation of the unemployed in paid informal work: a critical evaluation of the deterrence approach', *Environment and Planning C* 19,5: 729–49.

—— (2002a) 'Beyond the commodity economy: the persistence of informal economic activities in rural England', *Geografiska Annaler B* 83,3: 179–89.

—— (2002b) 'Harnessing social capital: some lessons from rural England', *Local Government Studies*, 28, 3: 183–195.

—— and Windebank, J. (1993) 'Social and spatial inequalities in the informal economy: some evidence from the European Community', *Area* 25,4: 358–64.

—— (1995a) 'Social polarisation of households in contemporary Britain: a "whole economy" perspective', *Regional Studies* 29,8: 727–32.

—— (1995b) 'Black market work in the European Community: peripheral work for peripheral localities?', *International Journal of Urban and Regional Research* 19,1: 23–39.

—— (1998a) *Informal Employment in the Advanced Economies: implications for work and welfare*, London: Routledge.

—— (1998b) 'The unemployed and informal sector in Europe's cities and regions', in P. Lawless, R. Martin and S. Hardy (eds) *Unemployment and Social Exclusion: landscapes of labour inequality*, London: Jessica Kingsley.

—— (1999a) 'The formalisation of work thesis: a critical evaluation', *Futures* 31,6: 547–58.

—— (1999b) *A Helping Hand: harnessing self-help to combat social exclusion*, York: York Publishing Services.

—— (2000a) 'Self-help and mutual aid in deprived urban neighbourhoods: some lessons from Southampton', *Urban Studies* 37,1: 127–47.

—— (2000b) 'The growth of urban informal economies', in R. Paddison (ed.) *Handbook of Urban Studies*, London: Sage.

—— (2001a) 'Reconceptualising paid informal exchange: some lessons from English cities', *Environment and Planning A* 33,1: 121–40

—— (2001b) *Revitalising Deprived Urban Neighbourhoods: an assisted self-help approach*, Aldershot: Ashgate.

—— (2001c) 'Beyond profit-motivated exchange: some lessons from the study of paid informal work', *European Urban and Regional Studies* 8,1: 43–56.

—— (2001d) 'Beyond social inclusion through employment: harnessing mutual aid as a complementary social inclusion policy', *Policy and Politics* 29,1: 15–28.

Williams, C.C., Aldridge, T., Lee, R., Leyshon, A., Thrift, N. and Tooke, J. (2001) *Bridges into work? An evaluation of Local Exchange and Trading Schemes (LETS)*, Bristol: The Policy Press.

Windebank, J. (1991) *The Informal Economy in France*, Aldershot: Avebury.

—— (1996) 'To what extent can social policy challenge the dominant ideology of mothering? A cross-national comparison of France, Sweden and Britain', *Journal of European Social Policy* 6: 147–61.

—— (1999) 'Political motherhood and the everyday experience of mothering: a comparison of child care strategies of French and British working mothers', *Journal of Social Policy* 28, 1: 1–25.

Wolfe, A. (1989) *Whose Keeper?*, Berkeley: University of California Press.

Woodward, R. (1995) 'Approaches towards the study of social polarisation in the UK', *Progress in Human Geography* 19: 75–89.

Wright, C. (1997) *The Sufficient Community: putting people first*, Dartington: Green Books.

Wuthnow, R. (1997) *Sharing the Journey*, New York: Free Press.

Yankelovich, D. (1995) *Young Adult Europe*, Paris: Yankelovich Monitor.

Young, M. and Wilmott, P. (1975) *The Symmetrical Family: a study of work and leisure in the London region*, Harmondsworth: Penguin.

Zafirovski, M. (1999) 'Probing into the social layers of entrepreneurship: outlines of the sociology of enterprise', *Entrepreneurship and Regional Development* 11: 351–71.

Zelizer, V.A. (1994) *The Social Meaning of Money*, New York: Basic Books.

Zimmerman, M.E. (1987) 'Feminism, deep ecology and environmental ethics', *Environmental Ethics*, 9: 21–44.

Zoll, R. (1989) *Nicht So Wie Unsere Eltern*, Oplandan: Westdeutscher Verlag.

Index